DENVER PUBLIC SCHOOLS
3 7137 04231415 0 793.74 LON
Wacky word problems : games

D1443668

Wacky Word Problems

Also in the Magical Math series

Dazzling Division

Delightful Decimals and Perfect Percents

Fabulous Fractions

Great Graphs

Groovy Geometry

Marvelous Multiplication

Measurement Mania

Magical Math

Wacky Word Problems

Games and Activities That Make Math Easy and Fun

Lynette Long

JOSSEY-BASS
A Wiley Imprint
www.josseybass.com

Copyright © 2005 by Lynette Long. All rights reserved
Illustrations copyright © 2005 by Tina Cash-Walsh. All rights reserved

Published by Jossey-Bass
A Wiley Imprint
989 Market Street, San Francisco, CA 94103-1741 www.josseybass.com

Published simultaneously in Canada

Design and composition by Navta Associates, Inc.

No part of this publication may be reproduced, stored in a retrieval system, or transmitted in any form or by any means, electronic, mechanical, photocopying, recording, scanning, or otherwise, except as permitted under Section 107 or 108 of the 1976 United States Copyright Act, without either the prior written permission of the publisher, or authorization through payment of the appropriate per-copy fee to the Copyright Clearance Center, Inc., 222 Rosewood Drive, Danvers, MA 01923, 978-750-8400, fax 978-646-8600, or on the Web at www.copyright.com. Requests to the publisher for permission should be addressed to the Permissions Department, John Wiley & Sons, Inc., 111 River Street, Hoboken, NJ 07030, 201-748-6011, fax 201-748-6008, or online at http://www.wiley.com/go/permissions.

Limit of Liability/Disclaimer of Warranty: While the publisher and author have used their best efforts in preparing this book, they make no representations or warranties with respect to the accuracy or completeness of the contents of this book and specifically disclaim any implied warranties of merchantability or fitness for a particular purpose. No warranty may be created or extended by sales representatives or written sales materials. The advice and strategies contained herein may not be suitable for your situation. You should consult with a professional where appropriate. Neither the publisher nor author shall be liable for any loss of profit or any other commercial damages, including but not limited to special, incidental, consequential, or other damages.

The publisher and the author have made every reasonable effort to ensure that the experiments and activities in this book are safe when conducted as instructed but assume no responsibility for any damage caused or sustained while performing the experiments or activities in the book. Parents, guardians, and/or teachers should supervise young readers who undertake the experiments and activities in this book.

Readers should be aware that Internet Web sites offered as citations and/or sources for further information may have changed or disappeared between the time this was written and when it is read.

Jossey-Bass books and products are available through most bookstores. To contact Jossey-Bass directly call our Customer Care Department within the U.S. at 800-956-7739, outside the U.S. at 317-572-3986, or fax 317-572-4002.

Jossey-Bass also publishes its books in a variety of electronic formats. Some content that appears in print may not be available in electronic books.

Library of Congress Cataloging-in-Publication Data

Long, Lynette.
Wacky word problems : games and activities that make math easy and fun / Lynette Long.
p. cm. —(Magical math)
Includes index.
ISBN 0-471-21061-7 (pbk. : alk. paper)
1. Mathematical recreations. I. Title.
QA95.L825 2005
793.74—dc22

2004014921

Printed in the United States of America
FIRST EDITION
PB Printing 10 9 8 7 6 5 4 3 2

Contents

1

The Magic of Word Problems

When you're not in class, how often do you ask yourself, "What is 821 + 53 + 444?" or "What is 16 × 4?" or "What is 32 divided by 8?" Probably not very often, unless you're reading your homework out loud. So why study math? Well, although you may not realize it, you answer these kinds of questions in word problems all the time.

For example, your parents are driving you to your favorite amusement park. You sit in the back seat, wondering, *How long until we get there?* If your parents tell you the distance to the park and how fast the car is going, can you figure it out? This is a word problem!

You stop at a restaurant after school. You look at the money in your pocket and wonder, *Do I have enough to get a meal and dessert?* You know the prices from the menu. Can you figure it out? Another word problem!

You get your spelling test back. Your score is 17 out of 21. You ask yourself, *What percentage did I get correct?* You know you'll get a B or better if you got more than 80 percent right. Can you figure it out? Yep, that's a word problem.

You go clothes shopping, and your favorite store has a sale on all jeans—30 percent off. You wonder, *How much will those jeans really cost me?* You know the original price of the jeans. Can you figure out the sale price? Another word problem.

You will be surprised at how many word problems you try to solve every day—problems about time, distance, money, percentages, and measurement, to name a few. In fact, word problems can be about anything, as long as they are written as a story and pose a mathematical problem.

Some students think word problems are difficult, because they pose a problem but don't tell you how to solve it. Word problems don't tell you to add these three numbers or multiply these two numbers. They just ask a question and give you the facts you need to solve it. The trick is to know what to do with those facts. That's what you're about to find out!

In this book you'll practice until you become a word problem master. You'll learn what to do when solving the most common types of word problems. It's not as hard as you think. There are clues to look for, rules to learn, and a few formulas to memorize, but after that, the word problems you encounter every day will seem like a breeze.

Why not get started? A lot of fun activities await you.

II

COMPUTATION

The word problems in this section use simple computation. To solve these problems, you will add, subtract, multiply, or divide. Learning how to solve these basic word problems is the first step in becoming a word problem master.

In this section you will play a card game as you learn to identify word cues, make up word problems about yourself, use the phone book to play a word problem game, and express your creativity by creating word problems to match mathematical expressions. Word problems are not just about math; they are also about words and the special way words are used and interpreted in word problems.

If you double the number of my age and subtract 2 you get 20. How old am I?

1

Word Clues

Learn what key words to look for in word problems while you play a card game.

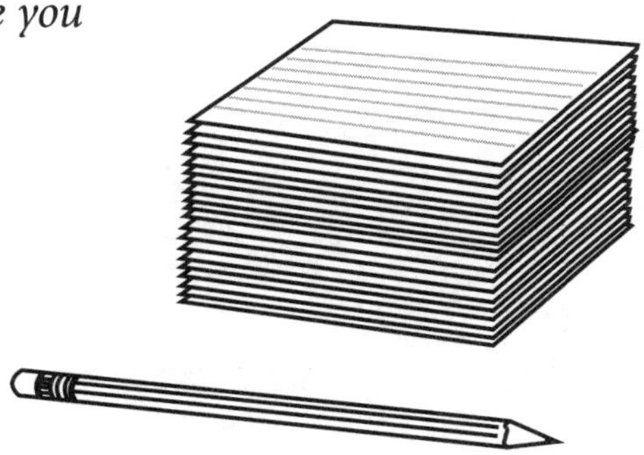

MATERIALS

pencil
index cards

Game Preparation

Write each of the following words on a different index card. Each of these words is commonly found in word problems, and every word provides a clue about how to solve the word problem. Some words mean you should use addition, some mean that you use subtraction, some mean that you use multiplication, and some mean that you use division to solve the word problem.

sum	difference	times	split
perimeter	less	area	divide
total	fewer	product	average
all	change	percent	share
altogether	minus	of	part

Game Rules

1. Deal each player three cards. Place the rest of the cards face down in the center of the table.
2. Players take turns picking a single card from the top of the pile and discarding a single card. The object of the game is to collect three cards that all mean addition, subtraction, multiplication, or division.
3. The first player to collect three cards that mean the same mathematical operation wins the round. The first player to win three rounds wins the game.

Tips and Tricks

Here's how to group the cards according to operation:

Addition	Subtraction	Multiplication	Division
Sum	Difference	Times	Split
Perimeter	Less	Area	Divide
Total	Fewer	Product	Average
All	Change	Percent	Share
Altogether	Minus	Of	Part

Wacky Word Problems

Read each of the following word problems and decide whether you have to add, subtract, multiply, or divide. Circle the correct answer. Underline the word that gave you the clue.

- If you ate 174 earwax flavor jelly beans and 63 dirt flavor jelly beans, how many more earwax jelly beans than dirt jelly beans did you eat? To solve this problem, do you add, subtract, multiply, or divide?

- If you scored 8 goals in each of 10 foosball games, how many goals did you score altogether? To solve this problem, do you add, subtract, multiply, or divide?

- If 4 frisky ferrets destroyed 3 red balls and 4 green balls, how many balls did they destroy altogether? To solve this problem, do you add, subtract, multiply, or divide?

- If you divided a 12-slice pepperoni, pineapple, and pickle pizza equally among 4 hungry students, how many slices of pizza would each person get? To solve this problem, do you add, subtract, multiply, or divide?

(Answers: Subtract; Multiply; Add; Divide)

BRAIN Stretcher

Each player rolls a single die. The number rolled represents an arithmetic operation.

1 = addition

2 = subtraction

3 = multiplication

4 = division

5 = choice

6 = roll again

Each player has 5 minutes to write down as many words or phrases as possible that are used in word problems to represent the word that was rolled. For example, if you rolled "1," which represents "addition," you could write "altogether" on your list. The player who can think of the most words or phrases for his or her operation wins the game.

2

All about Me!

Learn to solve simple word problems by working backward from answers to some questions about you.

Materials

- paper
- pencil
- index cards
- tape
- poster board

Procedure

1. How old are you? Take half of your age. Now subtract 2 from your answer. What's your answer?

2. Write the following word problem on an index card:

"If you take half my age and subtract 2 from the answer, you get *(answer here)*. How old am I?"

3. Write your age on the other side of the card.

4. What is the number of the month in which you were born? January is 1, February is 2, and so on. Double the number of the month in which you were born and subtract 2. What number did you get?
5. Write the following word problem on an index card:

   "If you take the number that represents the month in which I was born and double it and then subtract 2, you get *(answer here)*. What month was I born in?"
6. Put the name of the month in which you were born on the back of the index card.
7. Make up similar questions and answers for the following facts about you. Write each question on the front of an index card. Write the answer on the back.
   a. How tall am I in inches?
   b. What day of the month was I born?
   c. What size shoes do I wear?
   d. What is my zip code?
   e. How many students are in my class?
   f. What is the number of my classroom?
8. Now tape all the index cards on a sheet of poster board. Put the tape only at the top of the card so that you can lift it up. The questions should be face up so that if you lift up the card, the answer is on the other side.
9. Put a title at the top of the poster board.
10. Share your poster board with your family and friends. Can they figure out the facts about you?

3

All Four

Play this game using advertisements from a phone book to design word problems using four different operations.

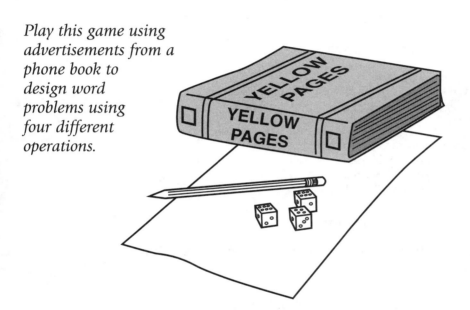

MATERIALS

3 dice
pencil
paper
phone book with advertisements
2 players

Game Rules

1. One player rolls all three dice. The numbers on the dice are arranged to form different numbers.

EXAMPLE

If 2, 3, and 6 are rolled, a list is made of the possible page numbers that can be formed by these three numbers.

236, 263, 326, 362, 623, 632

2. Players look at these page numbers in the phone book, and each player picks one company's advertisement. (Note: Players cannot pick the same advertisement.)
3. Each player writes four word problems about the items described in the advertisement. Each problem should be based on one of the four operations: addition, subtraction, multiplication, or division.

EXAMPLE

You find this advertisement in a phone book:

Mike's Boathouse
On the Banks of the Historic Potomac River
Rentals of Kayaks, Canoes, and Rowboats
Hourly/Daily Rates

Boat Rental Costs
Kayak—$6.50 per hour
Canoe—$4 per hour
Rowboat—$7 per hour

You then make up the following questions:

Which costs more to rent per hour, a kayak or a canoe? How much more? (subtraction)

How much would it cost to rent a rowboat for 4 hours? (multiplication)

How much would it cost to rent one rowboat, one canoe, and one kayak for 1 hour? (addition)

If it costs $10 to rent a canoe for half of a day, and a half-day is 4 hours, how much does it cost to rent the canoe per hour? (division)

4. Players then exchange the problems they've made up and solve each other's problems.

(Answers: Kayak, $2.50 more; $28; $17.50; $2.50)

4

Two Steppers

Play this game to construct word problems that take two steps to solve.

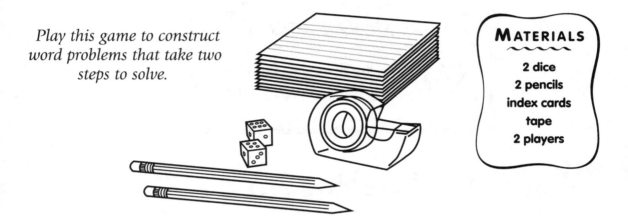

MATERIALS

2 dice
2 pencils
index cards
tape
2 players

Procedure

1. Each player rolls both dice. The numbers on the dice represent different operations.

1 = add

2 = subtract

3 = multiply

4 = divide

5 = choice

6 = roll again

2. Each player writes a number sentence on the index card, using the two operations rolled and whatever numbers he or she chooses.

For example, if "add" and "multiply" are rolled, the player might write, $(3 + 2) \times 6$ or $1 + (5 \times 7)$

3. Players exchange problems, and both players write word problems that match the expressions. For example, John gave 3 red lifesavers and 2 yellow lifesavers to each of 6 friends. Or Georgette had 1 white bead and 5 beads of each of 7 different colors. How many beads did she have altogether?

4. Players exchange problems again and solve them.

BRAIN Stretcher

What is the difference between $(3 + 2) \times 6$ and $3 + (2 \times 6)$?

Wacky Word Problem

If 5 children each caught 3 stink bugs before dinner and 6 slugs after dinner, how many critters did the children catch altogether?

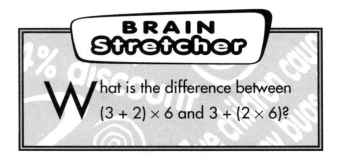

(Answer: 45)

III

MEASUREMENT

Measurement involves calculating distance, weight, volume, and time. Measurement word problems often ask you to convert from one measurement to another. What makes measurement complicated is that in the United States, both metric and English units are used. Not only do you have

to learn to convert between the different English units (for example, from pounds to ounces and vice versa), but you also must learn to convert between English and metric units (for example, between miles and kilometers). If that isn't enough, you have to learn to convert between different metric measurements; for example, how do you change from grams to kilograms?

In this section, you'll make conversion cards that will help you on your metric journey. You'll also enter into crazy physical contests with friends, research an African animal, make a metric bookmark, learn a dozen ways to make a quart, and play Measurement Jeopardy.

5

Conversion Cards

In this activity, you'll make a set of conversion cards to help you convert from one English unit of measurement to another.

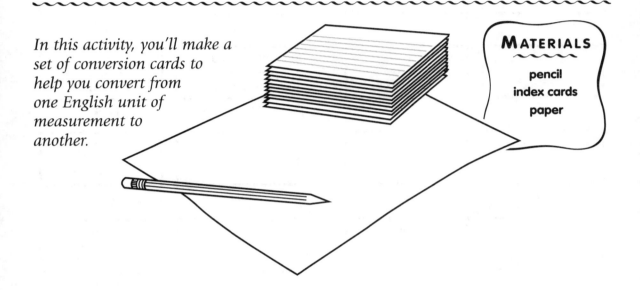

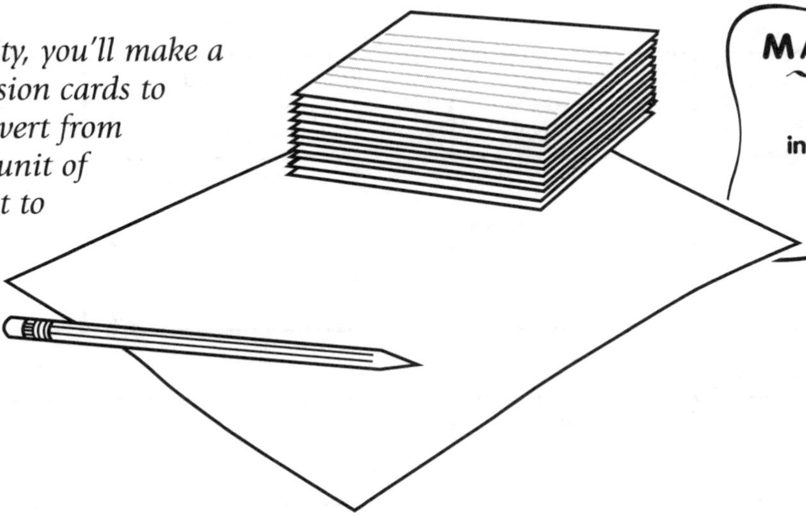

MATERIALS

pencil
index cards
paper

Procedure

1. Copy one of the following ratios on an index card. Copy its inverse (opposite) on the back of the same index card. You've just made your first conversion card.

$$\frac{\text{12 inches}}{\text{1 foot}}$$

front

$$\frac{\text{1 foot}}{\text{12 inches}}$$

back

2. Copy each of the following ratios on the front and the back of different index cards to create additional conversion cards.

3 feet / 1 yard	1 yard / 3 feet
front	*back*

2 pints / 1 quart	1 quart / 2 pints
front	*back*

5,280 feet / 1 mile	1 mile / 5,280 feet
front	*back*

4 quarts / 1 gallon	1 gallon / 4 quarts
front	*back*

16 ounces / 1 pound	1 pound / 16 ounces
front	*back*

60 seconds / 1 minute	1 minute / 60 seconds
front	*back*

2,000 pounds / 1 ton	1 ton / 2,000 pounds
front	*back*

60 minutes / 1 hour	1 hour / 60 minutes
front	*back*

1 tablespoon / 3 teaspoons	3 teaspoons / 1 tablespoon
front	*back*

24 hours / 1 day	1 day / 24 hours
front	*back*

2 cups / 1 pint	1 pint / 2 cups
front	*back*

3. Use the cards to change problems from one unit to another by following these three simple steps.

- Find a ratio that contains both the unit you want and the unit you want to change to. Place the ratio so that the unit you want to change to is on top and the unit you have is on the bottom.

- Change your original number to a ratio by placing it over a 1.

- Multiply the two ratios together. Cross out the units that are the same.

EXAMPLE

Suppose you want to change 8 quarts to gallons.

First, pick the ratio

$$\frac{1 \text{ gallon}}{4 \text{ quarts}}$$

since it contains both quarts and gallons. Notice that the quarts are on the bottom of the ratio and the gallon is at the top of the ratio.

Second, change the original number to a ratio by placing it over 1, so 8 quarts becomes

$$\frac{8 \text{ quarts}}{1}$$

Third, multiply the two ratios together. Since the word *quart* is on both the top and bottom of the equation, they cancel each other, leaving only gallons on top.

$$\frac{8 \text{ quarts}}{1} \times \frac{1 \text{ gallon}}{4 \text{ quarts}} = \frac{8 \text{ gallons}}{4} = 2 \text{ gallons}$$

Tips and Tricks

How do you multiply one ratio by another?

It's just like multiplying fractions. First multiply the numerators, next multiply the denominators. Finally, simplify.

Wacky Word Problems

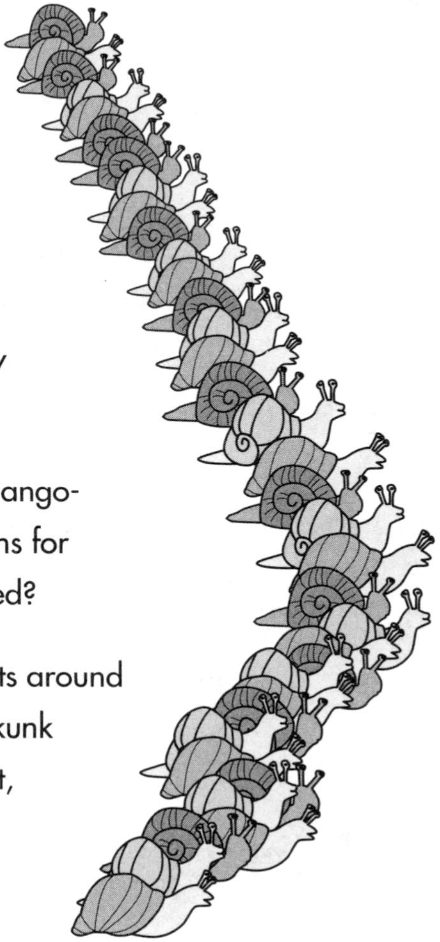

- Two quarts of wiggly earthworms were divided into cups. How many cups of earthworms were there?

- Thirty-six snails were lined up in a row. If each snail was 1 inch long, how many yards of snails were there?

- Fifteen tablespoons of fresh-squeezed mango-papaya juice were poured into teaspoons for tasting. How many teaspoons were tasted?

- Gardeners planted skunk cabbage plants around a park that was 3 miles around. If the skunk cabbage plants are planted 1 foot apart, how many skunk cabbage plants have been planted?

(Answers: 8 cups; 1 yard; 45 teaspoons; 15,840)

6

Crazy Contests

Learn to make measurement comparisons while having some crazy contests with friends.

MATERIALS

measuring spoons
bowl of water
timer
2 players

Procedure

There are three contests, each for two players. Players should take turns doing each part of each contest.

CONTEST 1

1. Get a set of measuring spoons, a bowl of tap water, and a timer.
2. Player 1 uses the tablespoon to drink as much water as possible in 15 seconds.

3. Player 2 uses a teaspoon to drink as much water as possible in 30 seconds.
4. Count the number of spoons the players can drink in the given time, then calculate the answer to the following word problem: If both players drank water at the same speed for 1 minute, who would drink the most water?

CONTEST 2

1. Player 1 runs backward 60 feet.
2. Player 2 hops forward 10 yards.
3. Time each player, and then calculate the answer to the following word problem: If both players were to keep moving at this pace for 100 yards, who would move the fastest?

CONTEST 3

1. Player 1 rolls a single die. For whatever number is rolled, Player 1 recites that times table as many times as possible in 60 seconds. For example, if the number 4 is rolled, Player 1 recites the four times table,
   $4 \times 1 = 4$, $4 \times 2 = 8$, $4 \times 3 = 12$, and so on.
2. Player 2 rolls a single die and recites the times table for the number rolled as many times as possible in 40 seconds.
3. Count how many times each person recites the times table and then calculate: At this same pace, how many times would each player be able to recite the times table he or she rolled in 5 minutes?

The player who wins two out of the three contests wins the game.

BRAIN Stretcher

Make up your own contests that involve measurement, and create word problems based on the contest results.

Wacky Word Problems

- Which is longer, a 42-inch hot dog or a 4-foot submarine sandwich?
- Which is longer, 2 miles of candy rope or 2,000 yards of taffy?
- Which is lighter, 11 ounces of marshmallows or 2 pounds of chocolate ice cream?
- Which is more, 40 cups of green bug juice or 5 gallons of purple sports drink?

(Answers: A 4-foot submarine sandwich; 2 miles of candy rope; 11 ounces of marshmallows; 5 gallons of purple sports drink)

7

Vital Statistics

Try this activity to learn to do conversions while you find out about some interesting African animals.

MATERIALS

computer with Internet access
printer
pencil
paper

Procedure

1. Pick an animal from the following list that you would like to research.
   - elephant
   - giraffe
   - chimpanzee
   - warthog
   - cheetah
   - leopard
2. Use the Internet to look up the following facts about the animal.
   - weight in pounds
   - weight at birth in pounds
   - height in feet and inches
   - life span in years
   - top speed in miles per hour
3. Copy down the facts about the animal on a piece of paper.

4. Now figure out the following conversions:
   - weight in ounces
   - weight at birth in ounces
   - height in inches
   - life span in minutes
   - top speed in feet per second
5. Add these conversions to your fact sheet.

BRAIN Stretcher

Figure out what your animal's vital statistics would be in metric measurements. Are any of the statistics the same?

Tips and Tricks

To change the measurements from one English measurement to another, use the following conversion tips.

- 1 pound = 16 ounces. To change pounds to ounces, multiply by 16.
- 1 foot = 12 inches. To change feet to inches, multiply by 12.
- 1 year = 365 days, 1 day = 24 hours, 1 hour = 60 minutes. To change years to minutes, multiply the number of years by 365, then by 24, and finally by 60.
- To change miles per hour to feet per second, multiply the speed in miles per hour by 5,280 and divide by 3,600.

To change from English to metric, use these tips.

- 1 pound = .454 kg. To change from pounds to kilograms, multiply the number of pounds by .454.
- Time is the same in both the metric and the English system.
- To change miles per hour to kilometers per hour, just multiply by 1.61, and you're all set.

Wacky Word Problems

- If a fish rode a bike for 10 miles, how far did it ride in kilometers?
- How much would a 1,000-pound cat weigh in kilograms?
- How many liters is 20 gallons of pistachio pudding?
- If one paperclip weighs 1 gram, how many pounds does a chain of 15,000 paperclips weigh?
- If a gorilla ran in a 40-kilometer race, how many miles did it run?
- If you split a liter of green slime between yourself and four friends, how many ounces of green slime do you each get?

(Answers: 16.1 kilograms; 454 kilograms; 78.6 liters; 33 pounds; 24.4 miles; 8.45 ounces)

8

Up and Down

Make a bookmark that will help you solve metric conversion problems.

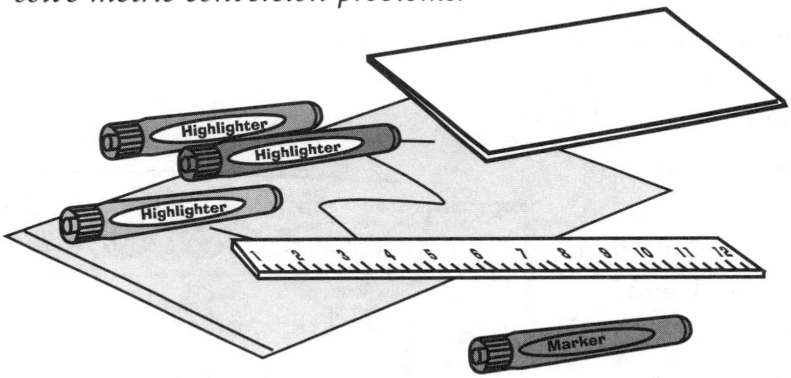

MATERIALS

4 × 6 index card
Magic Marker
ruler
3 highlighters
clear contact paper

Procedure

1. Cut the index card in half along the longest measurement. Now you should have two pieces, each 2 inches wide and 6 inches long.
2. Using the marker, divide the index card into eight rows that are each 3/4 of an inch wide.
3. Shade the rows using the highlighters. Alternate the colors so that your card has stripes.
4. Write each of the following words on a separate line of your bookmark.

    Metric Conversions

    kilometer

    hectometer

    dekameter

meter

decimeter

centimeter

millimeter

5. Turn your bookmark over. Draw a horizontal line 3 inches from the top of the card. Color the two sections of the card two different colors. Write the following directions in the top section of the card.

   Going up?

   Move the decimal one point to the left for each row up.

6. Write the following directions in the bottom half of the card.

   Going down?

   Add one zero for each place down.

7. Cover both sides of the bookmark with clear contact paper. Now you are ready to use your bookmark to mark your place in a book and help with conversions.

8. Use the bookmark to convert from one metric unit to another.

BRAIN Stretcher

Make bookmarks that show how to convert liters and grams.

EXAMPLE

Change 32 meters to centimeters.

Look at your bookmark.

You have to go down two rows to get from "meters" to "centimeters" on the bookmark.

Add two zeros to the end of 32.

32 meters = 3,200 centimeters

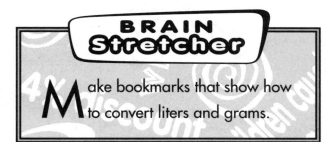

EXAMPLE

Change 450,000 centimeters to kilometers.

Look at your bookmark.

You have to go up five rows to get from "centimeters" to "kilometers" on your bookmark.

Move the decimal point five places to the left.

450,000 centimeters = 4.5 kilometers

Wacky Word Problem

Silly Sam hopped to school on his left foot. He hopped 100 times, and each of his hops was 1 meter long. How far away was his school in kilometers?

(Answer: 0.1 kilometer)

9

Quart Power

Learn how to solve problems involving units of measurement while using rice to figure out a dozen different ways to make a quart.

Materials

quart measuring cup
regular measuring cup
2 large bowls
2 bags of rice
paper
pencil
measuring spoons

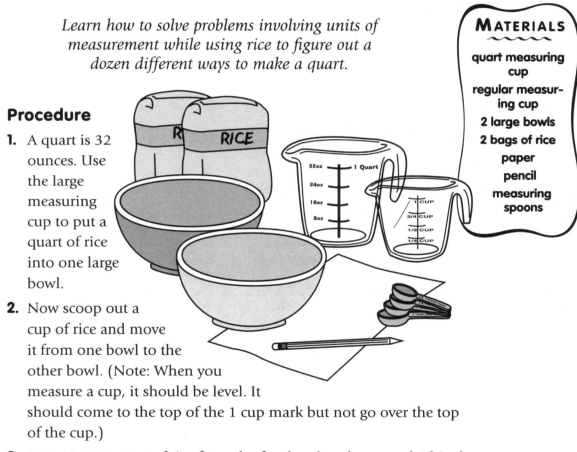

Procedure

1. A quart is 32 ounces. Use the large measuring cup to put a quart of rice into one large bowl.
2. Now scoop out a cup of rice and move it from one bowl to the other bowl. (Note: When you measure a cup, it should be level. It should come to the top of the 1 cup mark but not go over the top of the cup.)
3. Move 3 more cups of rice from the first bowl to the second. This shows one way to make a quart: 4 cups equal 1 quart.
4. On top of a sheet of paper, write, "Ways to make a quart." Write the numbers 1 to 12 down the left-hand side of the page.

5. Next to number 1, write 4 cups.
6. Now transfer 3 cups of rice from the second bowl back to the first bowl. Transfer the remaining cup back, using tablespoons. How many tablespoons did it take?
7. Next to the number 2, write 3 cups and ___ tablespoons.
8. Now figure out 10 other ways to make a quart. Use combinations of pints, quarts, cups, teaspoons, and tablespoons.
9. How many different ways do you think there are to make a quart?

BRAIN Stretcher

How many ways can you find to make a yard?

How many ways can you find to make a mile?

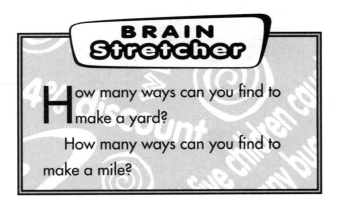

10

Measurement Jeopardy

Practice measurement conversions while playing Measurement Jeopardy.

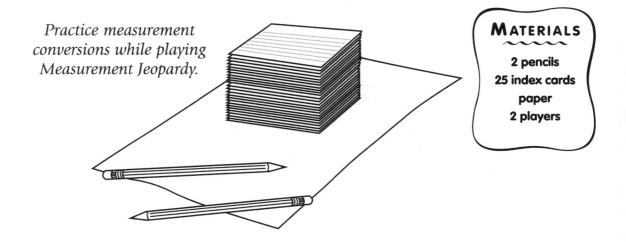

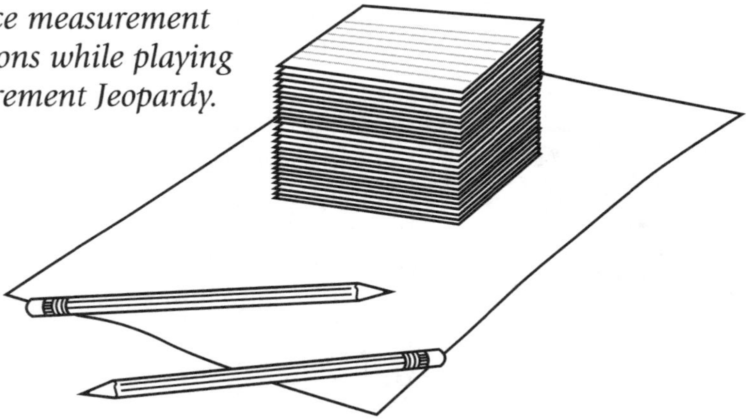

Materials

2 pencils
25 index cards
paper
2 players

Game Preparation

1. On the back of five index cards, write "inches."
2. On the back of five index cards, write "centimeters."
3. On the back of five index cards, write "meters."
4. On the back of five index cards, write "miles."
5. On the back of five index cards, write "kilometers."
6. Below the words on each set of cards, write one of the following numbers on each card: 10 20 30 40 50
7. Write a number between 1 and 100 on the other side of each card.
8. Lay the cards out on the table in columns of "inches," "centimeters," and so on. The cards should increase in value as you go down the column, so the first row is 10, the next 20, and so on.

inches 10	centimeters 10	meters 10	miles 10	kilometers 10
inches 20	centimeters 20	meters 20	miles 20	kilometers 20
inches 30	centimeters 30	meters 30	miles 30	kilometers 30
inches 40	centimeters 40	meters 40	miles 40	kilometers 40
inches 50	centimeters 50	meters 50	miles 50	kilometers 50

Game Rules

1. Player 1 chooses a card. The first player to make up a conversion problem, to which the answer is the number on the back of the card but is given in the units printed on the front of the card, wins the card.

EXAMPLE

If a "meters" card is turned over and the other side shows the number 3, a player might say, "What is 300 centimeters?"

2. Play continues until all the cards are gone from the table.
3. Players tally the points from the front of the cards they won.
4. The player with the most points wins the game.

BRAIN Stretcher

Make a second Measurement Jeopardy game using milligrams, grams, kilograms, ounces, and pounds.

IV

Counting and Logic Problems

Lots of word problems require you to use logic to solve them. They do not require a lot of mathematical skill, but they do require clear reasoning. Two of these kinds of problems are counting and logic problems. The easiest way to solve counting problems is to use Venn diagrams. The quickest way to solve logic problems is to reason them out.

In this section you will learn how to group things by attributes, use pennies to solve counting word problems, construct Venn diagrams while playing a matching game, and solve logic problems by drawing and arranging characters.

11

Block It Out

Try this activity to learn how to solve counting word problems by grouping things according to their attributes.

MATERIALS

3 pieces of string (each about 2 feet long)
index cards
colored pencils

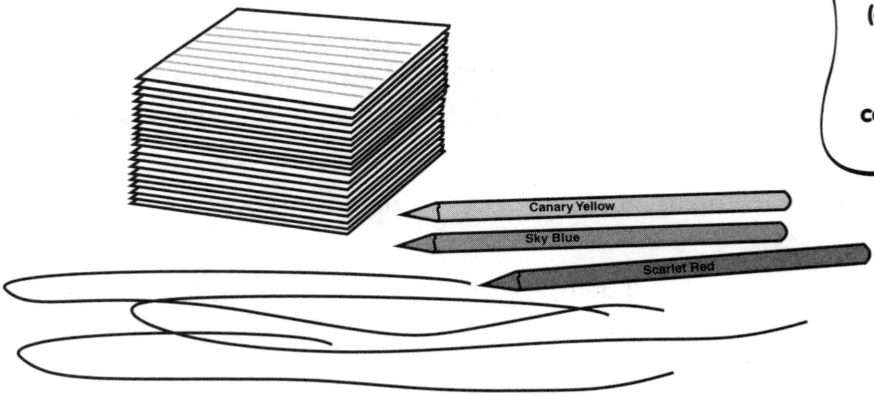

Procedure

1. Tie the ends of each of the three pieces of string together to make three circles.

2. Draw the following shapes on the index cards, one shape on each card.

small red circle	large blue circle
large red circle	small blue square
small red square	large blue square
large red square	small blue triangle
small red triangle	large blue triangle
large red triangle	small yellow circle
small blue circle	large yellow circle

small yellow square	small yellow triangle
large yellow square	large yellow triangle

3. Cut a piece of paper into eight sections. Write one of the following words on each small piece of paper.

large	blue
small	circle
red	square
yellow	triangle

4. Make two circles out of the strings, and place them on the table. Make the circles overlap. With these two circles, you have created four regions: the region in the first circle, the region in the second circle, the region in both circles, and the region outside both circles. You've just made what's called a Venn diagram.

5. Label each of the two circles with one of the eight words you wrote on the slips of paper.

6. Now place each of the shape drawings in the appropriate spot of the four regions you created.

EXAMPLE

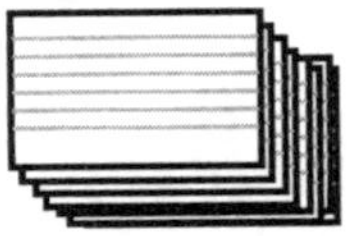

If you labeled the first string circle "red" and the second string circle "small," then you would place all of the red drawings in the first circle and all of the small drawings in the second circle, with the small red square, the small red circle, and the small red triangle in the region where the two string circles overlap. Place the remainder of the cards in a stack outside both circles.

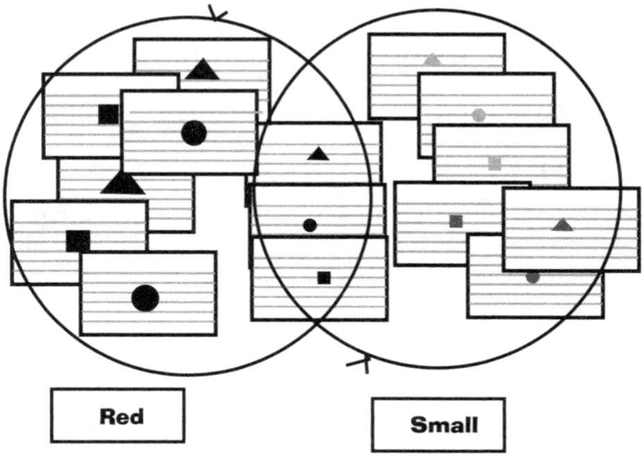

7. Label the two string circles with two different words and re-sort the piles.

BRAIN Stretcher

Make three circles out of string. Label each of the circles with a different word. Sort the colored drawings.

Wacky Word Problem

On a field trip to Pancake World, 5 kids had buckwheat pancakes and maple sugar sausage for breakfast, 3 had just pancakes, and a total of 10 had sausage. How many total kids were on the field trip? Draw a Venn diagram to solve this problem.

(Answer: 13)

12

Penny Math

Use pennies and string to solve counting word problems.

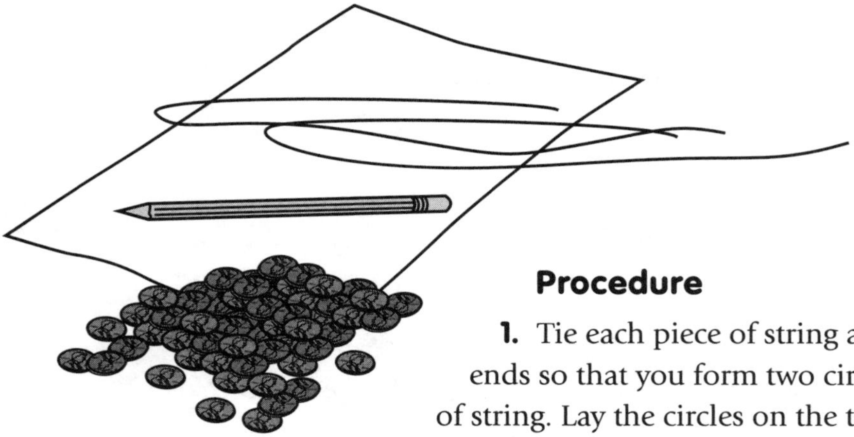

MATERIALS

2 pieces of string (each about 2 feet long)
50 pennies
paper
pencil

Procedure

1. Tie each piece of string at the ends so that you form two circles of string. Lay the circles on the table so that they overlap.
2. Use your pennies and string circles to solve this counting word problem.

 A total of 20 students signed up for the English or math classes.

 Ten students are taking English and math.

 Six students are taking English but not math.

 How many students are taking math but not English?

 How many students are taking math?

 To solve the problem:

 a. Use paper and a pencil to label one circle "math" and one circle "English."
 b. Place 10 pennies in the space where the two circles overlap. These 10 pennies represent the students who are taking both English and math.

c. Place 6 pennies in the "English" circle.

d. Count the total number of pennies you put down. Since 20 students, in total, are taking English, math, or both, and 16 students are accounted for, place 4 pennies in the other "math" section.

3. Now that the penny diagram is finished, you can answer the questions about the students.

Answer: Four students are taking math but not taking English.

The total number of students taking math is 14.

4. Use the pennies, the string, paper, and a pencil to solve the following problem.

There is a total of 25 students in the art and music classes.

Fifteen students take both art and music.

Eight students take music but not art.

How many take only art?

5. Try this problem.

Seventeen students play soccer, 6 play soccer and baseball, and 10 play baseball.

How many students are there altogether? Be careful.

(Answers: 2; 21)

BRAIN Stretcher

Put the two circles on the table. Overlap them. Put a total of 30 pennies in the three areas created by the circles. Now make up a statement that describes the way the pennies are arranged.

13

Venn Diagrams

Learn to use Venn diagrams while playing a matching game.

MATERIALS

- index cards
- 2 pencils
- timer
- 2 or more players

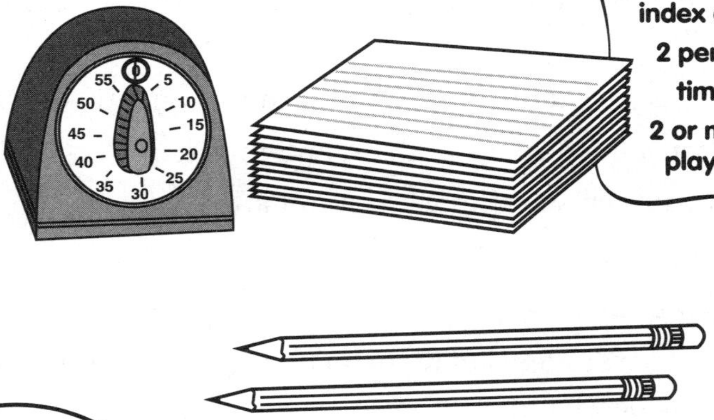

Game Rules

1. Each player gets six index cards. On three of the index cards, each player draws two overlapping circles to make a Venn diagram.

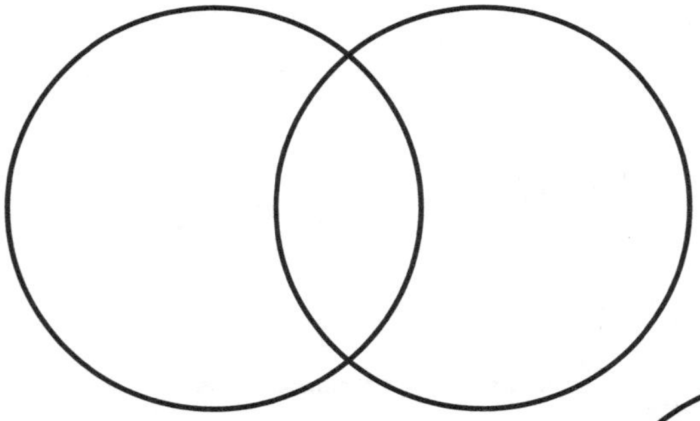

2. Each player writes one number in each of the three regions of the Venn diagram.

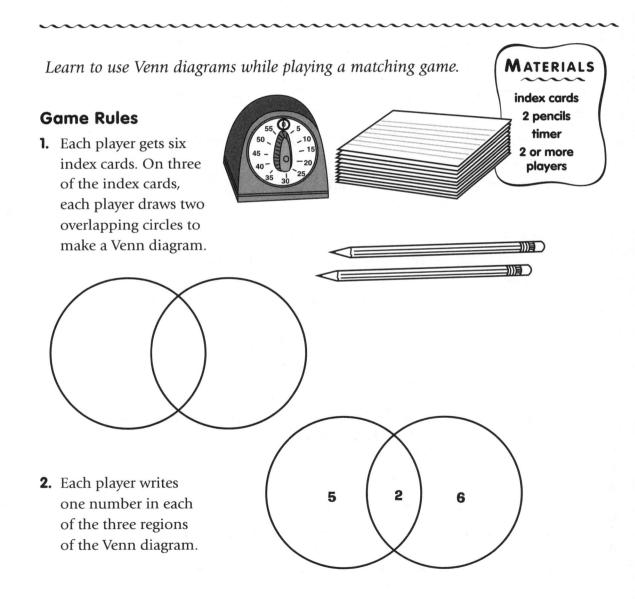

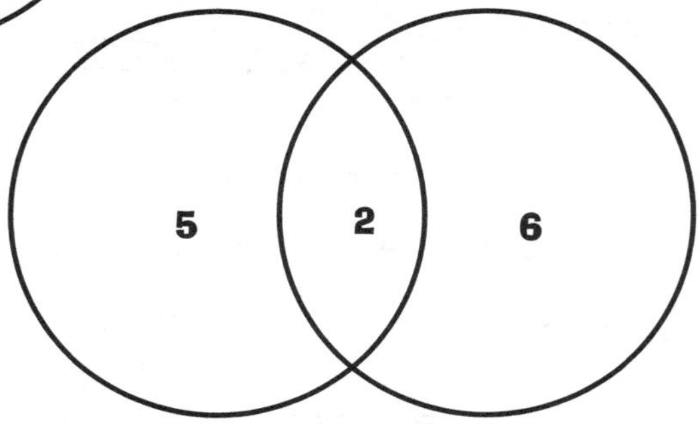

3. On separate index cards, players write a story to match each of the Venn diagrams they created. Each story should contain one question.

EXAMPLE

Five dogs have long hair, six dogs are small, and two small dogs have long hair. How many dogs are there altogether?

4. The timer is set for 5 minutes. Players pass their three cards to the left. Each player has until the time is up to figure out which card goes with which story and to answer the question on the card.

5. When the time is up, players take turns placing one diagram on the table, reading the story and the question, and providing the answer.

6. Next, each player takes four more index cards. On two of the index cards, each player draws three overlapping circles, creating a more complex Venn diagram. Seven separate regions are created inside the three circles.

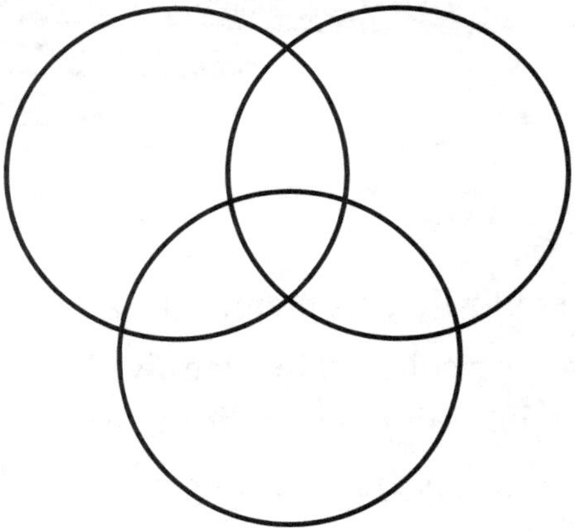

7. Each player writes numbers inside the seven regions of each circle. On separate index cards, players write a story and three questions about each of their Venn diagrams.

8. The timer is set for 5 minutes. Players pass their two cards to the right. Each player has 5 minutes to figure out which card goes with which story and answer the questions on the cards.

9. When the time is up, players take turns reading the stories, the questions, and the answers.

14

Who Is Older?

Have fun drawing characters and using clues about them to solve logic problems.

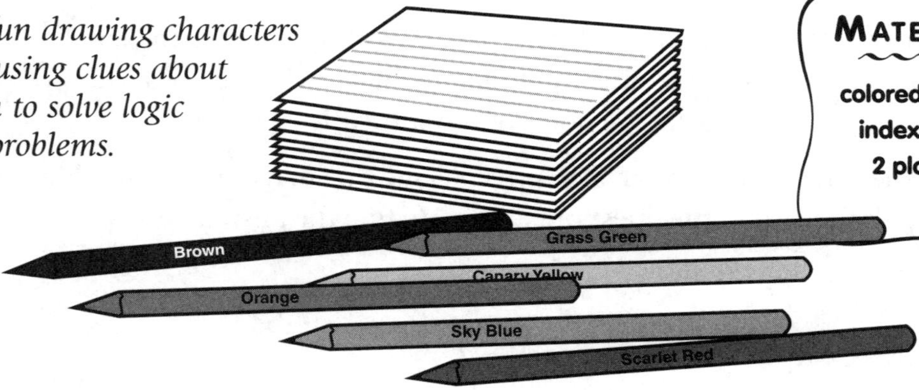

MATERIALS

colored pencils
index cards
2 players

Procedure

1. Read the following problem.

   There are four children in a family: Sue, Sam, Steve, and Sally. Figure out the order the children were born, using the following clues.

   Clue 1: Sue is Steve's younger sister.

   Clue 2: Sally is the oldest girl.

   Clue 3: Sam is older than Sally.

   Clue 4: Steve is older than Sam.

2. Draw each of the four children on an index card, and put that child's name under the drawing. Read the clues one at a time, and move the cards to match the clues.

   Clue 1: Sue is Steve's younger sister.

Put the card with Steve's picture's on the left, since Steve is older than Sue. Put the card with Sue's picture on the right.

Clue 2: Sally is the oldest girl.

Where do you put Sally? She could be older than Steve or between Steve and Sue. Put Sally under the two other cards, and read the next clue.

Clue 3: Sam is older than Sally.

Put Sam to the left of Sally, on the bottom row.

Clue 4: Steve is older than Sam.

Put Sam and Sally in between Steve and Sue.

Now your cards should be in the following order:

Steve, Sam, Sally, Sue

3. Read all four clues again, and make sure they all are true with the cards in this order.

4. Now solve this problem using four more cards. Figure out who is the tallest.

Clue 1: Kevin is shorter than Kim.

Clue 2: Ken is taller than Kevin.

Clue 3: Kris is the shortest.

Clue 4: Ken is shorter than Kim.

Game Rules

1. Each player writes three different names and draws pictures of these people on three index cards.

2. Players put their cards in any order. On a fourth index card, players write three clues about the order of the cards.

3. Players mix the cards and exchange cards and clues.

4. Players each try to put the other player's set of cards in the correct order. The first player to do this wins the first round.
5. Each player writes four different names and draws pictures of these people on four index cards.
6. Players put the cards in any order and write four clues about the order of the cards.
7. Players mix the cards and exchange cards and clues. The first player to put the four cards in the correct order wins the second round.
8. Each player writes five names and draws pictures of these people on index cards.
9. Players put the cards in any order, and each player writes five clues about the order of the cards.
10. Players mix the cards and exchange cards and clues. Who is the first player to put the five cards in the correct order?
11. The player who wins two out of three rounds wins the game.

V

Percentage Problems

Three types of percentage questions come up in word problems. In this section you'll learn how to solve all three types. The first type is taking the percentage of a number. The second type is finding the original number after you are given a number and a percentage. The third type is finding what percentage a certain number is of another number.

In this section you will play Percentage War, make a tipping wheel, take an imaginary shopping trip, make a sale chart, and play a banking game. It's all in fun, and you'll also learn a lot about solving percentage word problems.

15

Percentage War

Find out how to compute percentages while playing a fun game of Percentage War.

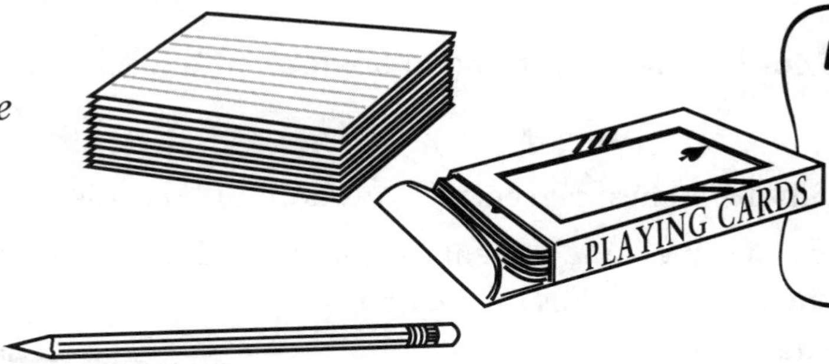

MATERIALS

pencil
index cards
playing cards
2 players

Game Preparation

Write each of the following words or phrases on a separate index card.

red	diamonds	less than 5
black	spades	queens
picture cards	clubs	seven or eight
ace	an odd number	
hearts	an even number	

Game Rules

1. Shuffle the index cards, and place them face down in the center of the table.
2. Deal each player 10 playing cards. Players place their sets of cards face up in front of them.
3. Each player draws 1 card from the stack of index cards. Each player uses the card that is drawn to fill in the following question: "What percentage of the cards in your hand is __________?"

EXAMPLE

"What percentage of the cards in your hand is *red*?"

4. Players answer the questions they draw as the questions apply to the cards they're holding.

EXAMPLE

If four of the cards in the player's hand are red, 40% would be the correct answer.

5. The player with the highest percentage wins both index cards.
6. If both players have the same percentage, players each draw a second index card. The player who has the highest percentage wins all four cards.
7. Players play a total of five rounds. The player with the most index cards at the end wins.

BRAIN Stretcher

Deal each player five cards, and play Percentage War again.

Tips and Tricks

To find out what percentage one number is of another number, just divide the first number by the second and multiply by 100.

Example

5 is what percentage of 20?

Divide: 5 divided by 20 is .25

Multiply: .25 times 100 = 25%

Wacky Word Problem

If 15 armadillos were born at the zoo and 3 of them were albino armadillos, what percentage of the armadillos born at the zoo were albino armadillos?

(Answer: 20%)

16

Tipping Wheel

Use your knowledge of figuring out percentages of whole numbers to make a tipping wheel.

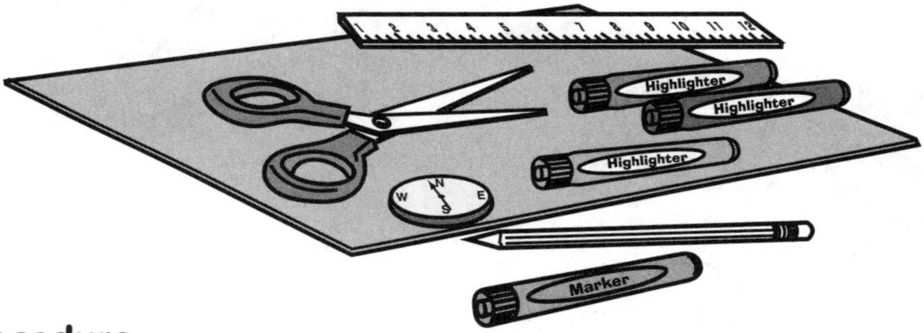

MATERIALS

compass
poster board
pencil
ruler
Magic Marker
highlighters
scissors

Procedure

1. Use the compass to draw a circle with an 8-inch diameter on the poster board. Inside the circle draw four smaller concentric circles. Trace these circles with a Magic Marker.
2. Use the ruler to divide the circle into eight equal sections, like the slices of a pizza.
3. In the outer ring, write the following amounts: $10, $15, $20, $25, $30, $35, $40, $45.
4. In the next ring, write numbers that are 10% of the numbers in the outer circle.
5. In the next ring, write numbers that are 15% of the numbers in the outer circle.
6. In the innermost ring, write numbers that are 20% of the numbers in the outer circle.

7. Write "Tips" in the center of the circle.

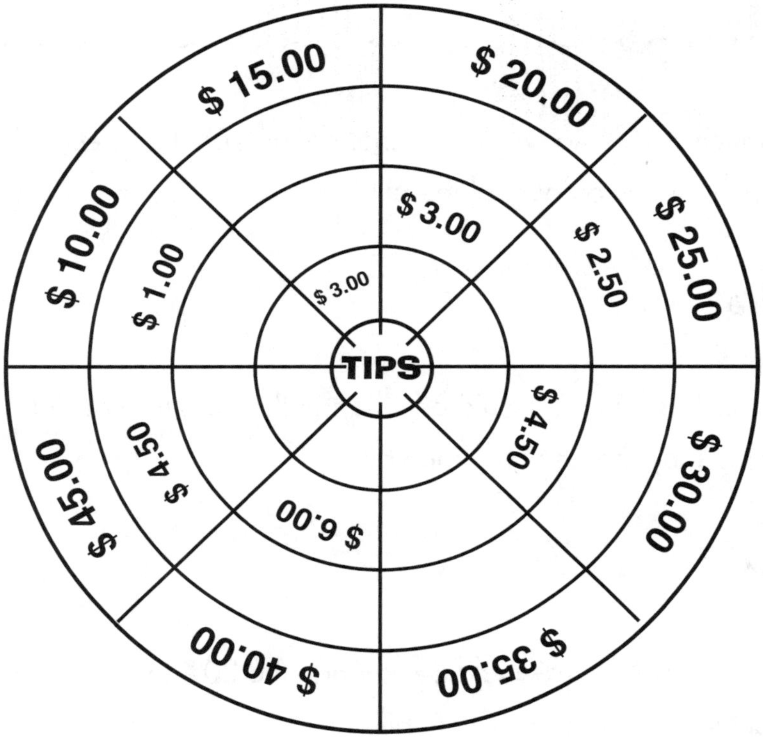

8. Parts of the tipping wheel have been filled in for you. Fill in the missing values, using the Tips and Tricks section for help.

9. Use highlighters to color the rings of the circle different colors.

10. Cut out the tipping wheel.

11. Use the tipping wheel at a restaurant to determine the amount of a tip. Look up the price of a meal in the outer circle. Round the bill up. Now look at the numbers in the pie slice. The closer they are toward the center of the circle, the better the tip. If the service was average, leave the middle amount. If the service was exceptional, leave the larger amount. If the service was poor, leave the lesser amount. Always round the tip amount to the nearest dollar when leaving a tip.

Tips and Tricks

To find out how much a 10% tip would be, change the 10% to a decimal (0.1) and multiply it by the amount of the bill.

Example

$0.1 \times \$32 = \$3.20$

Leave a $3 tip.

To find out how much a 15% tip would be, change the 15% to a decimal (0.15) and multiply it by the amount of the bill.

Example

$0.15 \times \$32 = \$4.80$

Leave a $5 tip.

To find out how much a 20% tip would be, change the 20% to a decimal (0.2) and multiply it by the amount of the bill.

Example

$0.2 \times \$32 = \$6.40$

Leave a $6 tip.

17

Sales Galore

Learn to take percentages of a number while you go on an imaginary shopping spree.

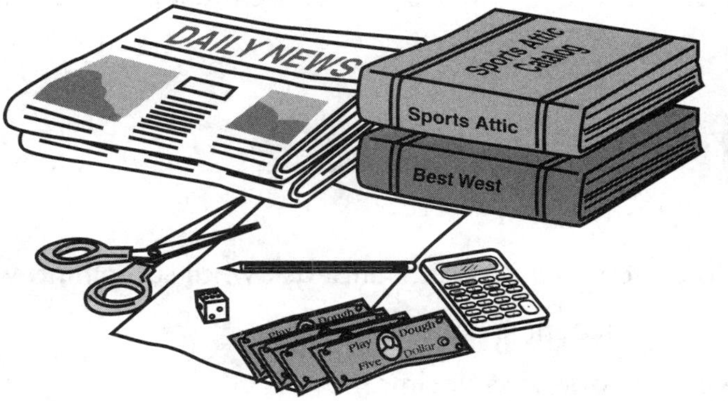

MATERIALS

scissors
old newspapers and catalogs
play money
die
pencil
paper
calculator
2 players

Game Preparation

Players cut pictures out of catalogs or newspapers of items they would like to buy. All the items selected should have the prices listed. The total cost of all the items selected by both players should be approximately $1,000. Items should vary in price from $250 to $10 or less.

Game Rules

1. Each player is given $300 in play money to spend.
2. Each player rolls the die. The player with the highest number shops first.
3. Player 1 rolls the die. The die tells the player what percentage to take off the purchase. If a player rolls a "1," the price is reduced 10%; if a 2 is rolled, take 20% off; for a 3, take 30% off; for a 4, take 40% off; for a 5, take 50% off, and for a 6, take 60% off the purchase price.
4. Once a player knows what discount applies for this purchase, the player selects one item to purchase at the discounted rate. The player picks up the picture, takes the discount, and pays for the item with play money.

5. Player 2 now rolls the die to determine the next discount, selects an item to purchase, and pays the sale price for the item.
6. Players continue to roll the die at each turn and use the new discount on each new purchase.
7. Once all the money is spent, players review their purchases. They can trade items if they like.
8. The winner is the player whose purchases total the most pre-sale value.

Tips and Tricks

It takes three steps to solve word problems that ask what something will cost after a percentage discount is figured in.

First, change the percentage to a decimal.

Next, multiply the decimal by the original price to find the dollar amount the item is discounted.

Finally, subtract the amount the product is discounted from the original price.

Example

If a $30 item is 20% off, how much will it cost on sale?

First, change 20% to a decimal.

$20\% = .2$

Next, multiply .2 times $30.

$.2 \times 30 = 6$

The item is discounted by $6.

Finally, subtract $6 from $30 to find the discounted price. ($24)

Wacky Word Problem

If the superpowered supersonic skateboard you wanted last year was originally $200 and is now 70% off, how much does it cost?

(Answer: $60)

18

How Much?

Make a chart to help you calculate sale prices and take it on shopping trips.

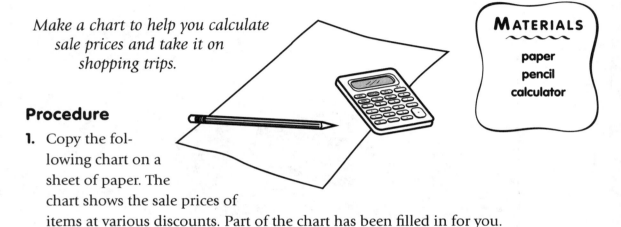

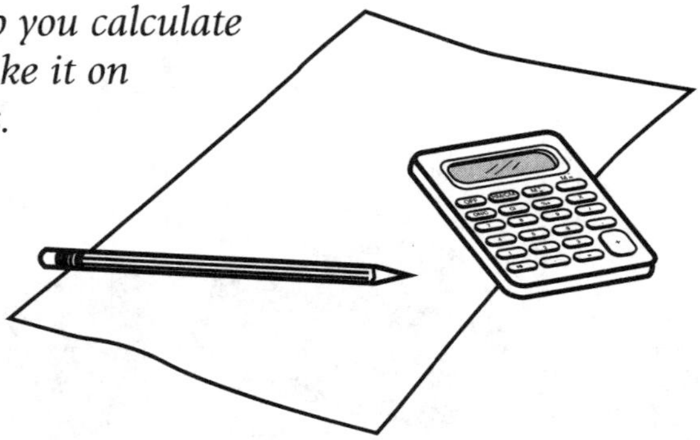

MATERIALS

paper
pencil
calculator

Procedure

1. Copy the following chart on a sheet of paper. The chart shows the sale prices of items at various discounts. Part of the chart has been filled in for you.
2. Fill in the rest of the chart. Start by filling in the first row. If something that originally cost $1 was listed as 10% off, how much would it cost after the discount? Change the percentage (10) to a decimal (.1). Multiply 1 by .1 and subtract the answer (.1) from 1. The answer is .9 or 90 cents. Keep going until the whole chart is filled in.
3. To read the chart, put the index finger of your left hand on the original price in the left-hand column and your right index finger on the discounted rate across the top of the chart.
4. Move your left finger across, and your right finger down until they meet. That is the discounted price.
5. Take the chart when you go shopping, to quickly figure out what things cost when they're on sale.

DISCOUNT

Original Cost	10%	20%	25%	33%	40%	50%	60%
$1	.90						
$5							
$10			$7.50				
$20							
$30							
$40							
$50	$45						
$60				$40			
$70							
$80							
$90							
$100						$50	

BRAIN Stretcher

How do you figure out the sale price of an item that's not on the chart? How much would a $250 TV cost if it were 25% off? How much would a $35 sweater cost if it were 20% off? How much would a $14 T-shirt cost if it were 20% off?

19

Simple Interest

Learn how to compute interest problems while playing a simple game.

MATERIALS

paper
pencil
playing cards
play money
die
calculator
2 to 4 players

Game Preparation

1. At the top of a single sheet of paper, write, "Years in Bank." Next, divide the rest of the sheet of paper into four columns. Write each player's name at the top of a column. Under each name, write the numbers 1 to 25 down the column. (The chart below shows just the first few rows.)

Years in Bank

Player 1	Player 2	Player 3	Player 4
1	1	1	1
2	2	2	2
3	3	3	3
4	4	4	4
5	5	5	5
6	6	6	6

2. Remove the kings, the queens, and the jacks from the deck of playing cards, and set them aside.
3. Divide a second sheet of paper into five sections. Write each player's name in a single section. Label the final section "The Bank." Place all the play money in the bank.

Game Rules

1. Place $100 of play money in each player's bank account. Place the rest of the money in "The Bank."
2. Player 1 rolls a single die to determine the percentage of interest his or her $100 will earn per year in the bank. If a 1 is rolled, the interest rate is 1%; if a 2 is rolled, the interest rate is 2%; and so on.
3. Player 1 now picks the top card from the deck to determine how long to leave the $100 in the bank at this interest rate.
4. Player 1 crosses out the same amount of numbers in his or her column on the "Years in Bank" sheet as shown on the card that was picked. For example, if Sue picks an ace, she crosses out the number 1. If she picks a 2, she crosses out the number 1 and the 2, and so on.
5. Player 1 determines how much interest is earned by keeping the $100 in the bank for that length of time at that interest rate and moves that much money from "The Bank" to his or her account.

EXAMPLE

A player rolls a 3 on the die and picks the 5 of clubs.

The player has invested $100. The player will earn 3 percent interest a year for 5 years or a total of $15. The player transfers $15 from the bank to his or her account.

6. Player 2 rolls the die to determine the annual interest rate.
7. Player 2 draws a playing card from the deck to determine how long to leave $100 in the bank at that interest rate.

8. Player 2 crosses out the numbers on the "Years in Bank" sheet that are equal to the number indicated on the card.
9. Player 2 calculates the interest and transfers that amount from the bank to his or her account.
10. Player 3 rolls the die, selects a card, and crosses out the number of years on the "Years in Bank" sheet that's equal to the number on the card. The player then computes the interest earned on the original investment and transfers that amount from the bank to his or her account.
11. Play continues until one player has kept money in the bank for a total of 25 years. At this time, all players count how much money they have. The player with the most money wins the game.

Tips and Tricks

"Simple interest" means you do not earn interest on the interest earned. You earn interest only on your original investment.

To compute the total interest paid is easy.

Just multiply the amount of the original investment by the interest rate by the years the money was invested.

Interest Earned = (Original Investment) (Interest Rate) (Years Invested)

Example

If you invest $100 at 2% simple interest for 4 years, how much interest will be earned?

Interest Earned = ($100) (.02) (4) = $8

Wacky Word Problem

If you're 10 years old, and you put the $100 your Aunt Bessie gave you for feeding her baby goats last summer into the bank now, and you earn 5% simple interest per year, will you have enough money when you are 21 to buy your own baby goat if each goat costs $150?

(Answer: Yes, you will have $155.)

VI

Distance Problems

Probably the most famous types of word problems are distance problems. In a distance problem, you figure out one of three things.

- How far away someplace is
- How long it takes to get somewhere
- How fast you are going

All distance problems use one formula. If you know this formula, you can be a distance whiz. All you have to do is change the formula around, and you can solve any distance problem.

In this section you will figure out how fast you and the members of your family walk, how far it is to different places in your neighborhood, how slowly a ball travels, how long it takes to go to different places in the United States, and how fast you can run. You'll also solve distance problems as you race toy cars. Ready, set, go!

20

How Fast?

Learn to calculate speed by figuring out how fast you and the members of your family walk.

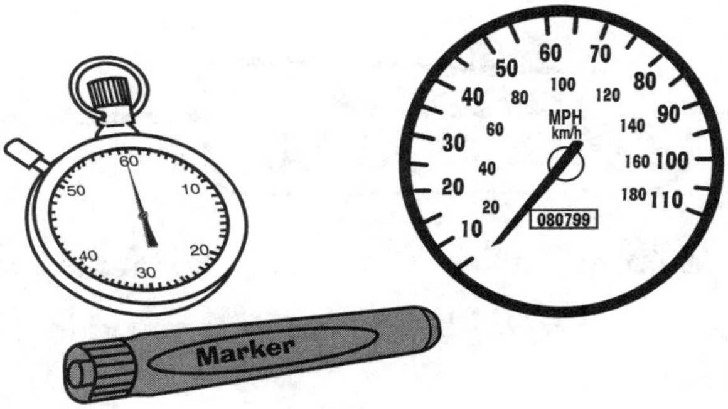

MATERIALS

- your family
- car odometer
- a marker of some sort
- a watch or a stopwatch

Procedure

1. Have one of your parents use the car odometer to measure a mile, starting at your house. Use something to mark the end of the mile.
2. Ask all the members of your family to walk this mile at their own pace and time how long in minutes it takes for them to walk this distance.
3. Look for each person's time on the following list to figure out how fast each person walked one mile. If you can't find the exact time, use the closest time.
   - 60 minutes = 1 mile per hour
   - 45 minutes = $1\frac{1}{2}$ miles per hour
   - 30 minutes = 2 miles per hour
   - 20 minutes = 3 miles per hour
   - 15 minutes = 4 miles per hour
   - 12 minutes = 5 miles per hour
   - 10 minutes = 6 miles per hour
   - $8\frac{1}{2}$ minutes = 7 miles per hour
   - 6 minutes = 10 miles per hour
   - 5 minutes = 12 miles per hour

4. Rate (Speed) is equal to Distance divided by Time. Use the rate formula to calculate each person's walking rate.

$R = D/T$

How do your answers compare to the rates you picked in Step 3?

For example, if you walked the mile in 20 minutes, your rate would be 1/20 (.05). To change this into miles per hour, multiply .05 by 60 to get 3.

BRAIN Stretcher

How fast can you bike 3 miles? Ride your bike on the same 1-mile course. Use $R = D/T$ to find out your speed in minutes, then convert it to miles per hour. Multiply your answer by 3 to see how long it would take you to bike 3 miles at this same pace.

SUPER BRAIN Stretcher

How long would it take each person to walk ⅓ of a mile? How about 3 miles? Use the rate and distance formula to figure it out.

Wacky Word Problem

How fast does a whippet run in miles per hour if it runs a half-mile in 1 minute?

(Answer: 30 miles per hour)

21

How Far?

Figure out how far it is to different places in your neighborhood using your own walking rate.

MATERIALS

pencil
paper
stopwatch
calculator

Procedure

1. Copy this chart on a piece of paper.

Places I Walk To	Walking Time in Minutes	Walking Time in Hours	Distance in Miles

2. Enter five places on the chart that you typically walk to.

3. Walk to each one, starting from your house. Walk at the same pace you used while walking a mile in the previous activity. Time how long it takes you to walk to each place. Enter the time in minutes in the chart.

4. Change each of the times in minutes to times in hours. To figure these out, just divide the number of minutes you walked by 60. The result is the number of hours or the part of an hour that you walked.

EXAMPLE

If you walked for 30 minutes, divide 30 by 60.
30 divided by 60 is ½ hour or .5 hours.

5. Use the distance formula to figure out how far it is to each of the places you usually walk to.

The Distance formula is Distance equals Rate multiplied by Time, or

$D = R \times T$

Tips and Tricks

To figure out the distance, just multiply the Rate (which is the speed you walk) by the Time (how long you walked).

Example:

If your walking speed is 4 miles per hour and you walked for 12 minutes (.2 hours) to get to a friend's house, what is the distance to your friend's house?

$4 \times .2$ or .8 miles

Wacky Word Problem

How far does a kangaroo have to hop to get to its favorite watering hole if the kangaroo hops at 6 miles per hour and reaches the watering hole in 20 minutes?

(Answer: 2 miles)

22

How Slow?

Learn how to solve speed problems while playing with a ball.

MATERIALS

large room
masking tape
tape measure
3 different types of balls (such as a tennis ball, a volleyball, a ping-pong ball, a basketball, or a soccer ball)
stopwatch
calculator
2 players

Procedure

1. Make a 10-foot track on the floor of a large room (at least 10 feet long). First, make a starting line by putting a strip of masking tape on the floor. Next, measure 10 feet and place another strip of masking tape on the floor to mark the finish line.

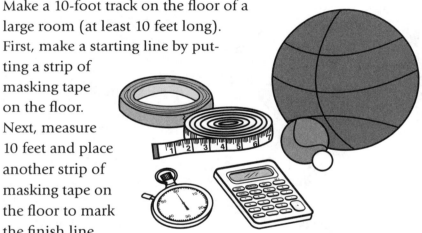

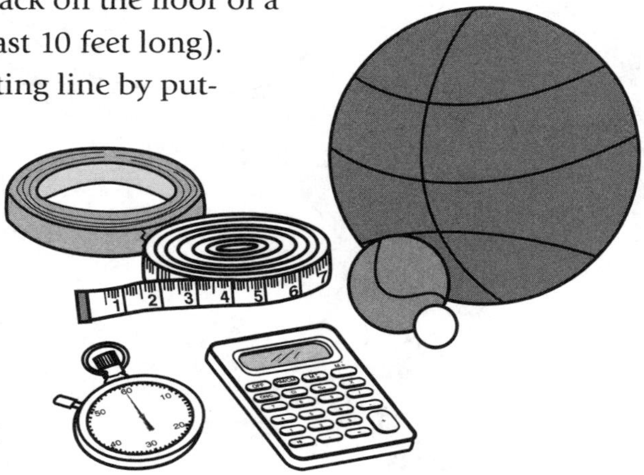

2. Sit on the floor behind the starting line, and roll a ball as slowly as you can across the finish line. If the ball does not reach the finish line, you lose your turn.
3. Use the stopwatch to time how long it takes for the ball to cross the finish line.
4. Compute the speed of the ball in feet per second. Just divide 10 by the number of seconds it took for the ball to travel 10 feet.

EXAMPLE

If it took 5 seconds for the ball to travel 10 feet, just divide 10 by 5. The answer is 2. The ball is traveling 2 feet per second.

5. Have a friend roll the ball as slowly as possible. Compute your friend's speed.
6. Each player gets a total of three chances to roll the ball. The player with the slowest speed wins the round.
7. Repeat the game with the second ball. Each player has three chances to roll the ball. The player with the slowest speed wins the round.
8. Repeat the game with the third ball. The player with the slowest speed wins the round.
9. The player who wins two out of the three rounds wins the game.

BRAIN Stretcher

How many feet per second can your favorite pitcher throw a ball?

On the Internet, look up pitching speeds of the top pitchers.

To change the speed from miles per hour to feet per second, first multiply the speed by 5,280 (that's the number of feet in a mile). Next, divide the answer by 3,600 (the number of seconds in an hour). The answer is the number of feet per second that a pitcher throws the ball.

23

How Long?

Calculate how long it would take for you walk, drive, and fly to different places in the United States.

MATERIALS

- one week's worth of mail
- pencil
- paper
- U.S. map or a computer with an Internet connection
- calculator

Procedure

1. Collect one week's worth of mail from your house (with your parents' permission). Look at the postmarks, and write down where each piece of mail originated on a chart like the following one. Try to pick mail from interesting places in the United States.
2. Use a U.S. map or the Internet to figure out approximately how many miles it is from your house to each of the places on your chart. Enter the results on the chart.

Origin of Mail	Miles from My House	Travel Time on Foot	Travel Time by Car	Travel Time by Plane

3. Assume that you walk at a rate of 3 miles per hour. How many hours would it take you to walk from your house to each of the cities on your chart? To find out, just divide the number of miles to each of the cities by 3. Enter the results on the chart.
4. Assume that a car goes an average of 50 miles per hour. How many hours would it take to drive from your house to each of the cities? To find out, just divide the number of miles to each of the cities by 50. Enter the results on the chart.
5. Assume that a plane goes an average of 500 miles per hour. How many hours would it take to fly from your house to each of the cities? To find out, just divide the number of miles to each of the cities by 500. Enter the results on the chart.

Tips and Tricks

The Time it takes to get somewhere is always the Distance divided by the Rate, or

$$T = D/R$$

Wacky Word Problem

A flying saucer can travel 5,000 miles per hour. How long will it take the aliens to get from New York to Los Angeles if these cities are 2,500 miles apart?

(Answer: $\frac{1}{2}$ an hour)

24

Ten-Second Sprints

Learn how fast you can run while you practice computing and converting rate problems.

MATERIALS

track or a large outdoor space
chalk
stopwatch
tape measure
pencil
paper
calculator
helper

Procedure

1. Use the chalk to mark a starting line on the ground.
2. Stand behind the line. When your helper clicks the stopwatch and yells, "Start," run as fast as you can.
3. After 10 seconds, when your helper yells, "Stop," stop running.
4. To figure out how far you ran, use the tape measure to measure the distance between the starting line and the mark where you stopped.
5. Make a chart like the following one, and enter the distance you ran in the 10-second sprint.

	Distance Run in Feet	Speed in Feet/Second	Speed in Feet/Hour	Speed in Miles/Hour
10-Second Sprint				
10-Second Uphill Sprint				
10-Second Downhill Sprint				

6. Now compute your speed. (Remember, $R = D/T$.) How far did you run in 10 seconds? Divide this by 10 to figure out how far you can run in 1 second. Enter the results on the chart.
7. Now compute how many feet you can run in 1 hour. First, multiply the number of feet you can run in a second by 60 to find out how many feet you can run in a minute. Next, multiply the answer by 60, since there are 60 minutes in an hour. Enter the result on the chart.
8. Now you can determine how many miles you can run in an hour. Since there are 5,280 feet in a mile, just divide the number of feet you ran by 5,280. Enter the result on the chart.
9. Repeat the activity on a course that goes uphill. How many feet did you run? How many feet per second can you run uphill? How many feet per hour can you run uphill? How many miles per hour can you run uphill? Enter all the results on the chart.
10. Repeat the activity on a track that goes downhill. How many feet did you run? How many feet per second can you run downhill? How many feet per hour can you run downhill? How many miles per hour can you run downhill? Enter all the results on the chart.

BRAIN Stretcher

How many miles per hour can you run backward? How many miles per hour can you skip?

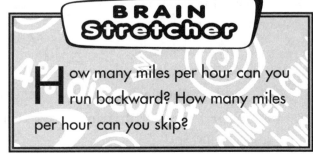

25

Racing Cars

Solve distance problems while you play a game with toy cars.

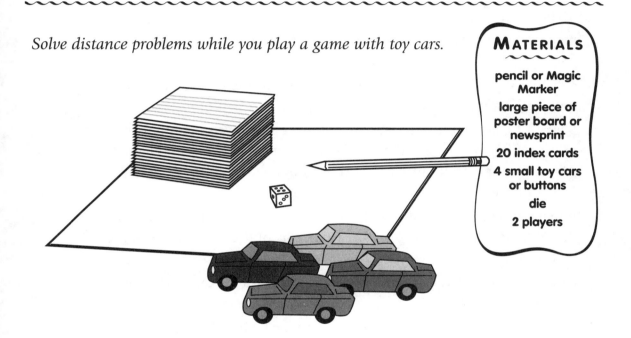

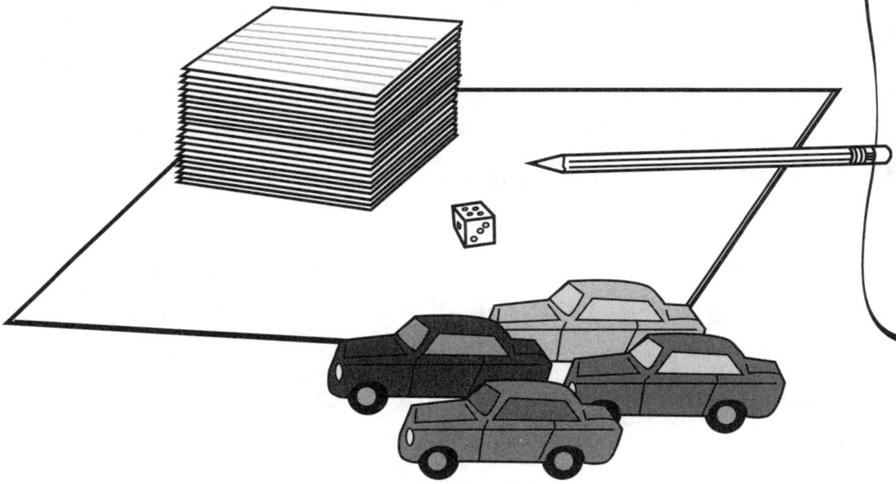

MATERIALS

pencil or Magic Marker

large piece of poster board or newsprint

20 index cards

4 small toy cars or buttons

die

2 players

Game Preparation

1. Draw a grid on the poster board or newsprint. The grid should be 30 squares tall and 30 squares wide. Each square represents 10 miles.
2. Mark the center of the grid with a large dot.
3. Write each of the following expressions on four index cards: 10 miles per hour, 20 miles per hour, 30 miles per hour, 40 miles per hour, and 50 miles per hour. In other words, you will end up with four cards that say "10 miles per hour," four cards that say "20 miles per hour," and so on, for a total of 20 cards.

Game Rules

1. Shuffle the index cards, and place them face down in the center of the table.
2. Each player chooses a car. Player 1 places a car in the lower left-hand corner of the board. Player 2 places a car in the lower right-hand corner of the board. The object of the game is to get your car nearest to the center of the board in five turns.
3. Player 1 picks the top index card. The card indicates how fast the car is going.
4. Now Player 1 rolls the die. The die indicates how many hours the car is traveling at a certain speed.
5. Player 1 figures out how far the car will move if it travels the number of hours rolled on the die at the speed shown on the drawn card. Player 1 moves the car that distance in one direction: up, down, left, or right. Remember, each square represents 10 miles.
6. If there is not enough room on the board for a player to move a car in any direction, the player must place the car in one of the four corners of the board.
7. Now Player 2 picks a card, rolls the die, and moves the car in any one direction.
8. No two cars can occupy the same space on the board. If one player's car already occupies a space, the second player must move his or her car in a different direction for that turn.
9. After each player has taken five turns, the player whose car is closest to the center wins the game.

VII

ALGEBRA

Algebra is a process of finding a missing number or numbers in a mathematical sentence. Letters of the alphabet, usually x, y, and z, are used to symbolize the missing numbers. Understanding algebra is the key to solving many different kinds of word problems.

In this section you will learn to create mathematical expressions by thinking up crazy answers to questions; solve simple algebraic expressions by using cups and pennies; play a guessing game using a magic shoebox; solve equations with two variables by using cups, plates, and pennies; and solve problems by using trial and error.

26

Crazy Sentences

In algebra, an expression is a group of numbers or symbols connected by operation symbols that represent a number, such as 3 + 2, or x – 5. In this activity, you'll learn to express numbers as crazy expressions.

MATERIALS

pencil
index cards
timer
2 or more players

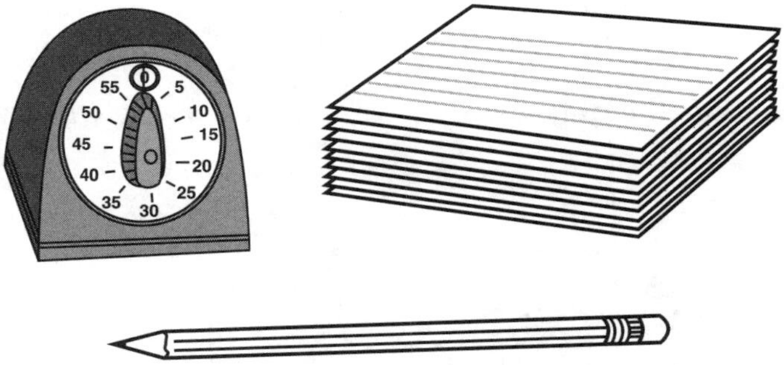

Game Preparation

1. Write each of the following sports questions on a separate index card.

- How many cards are in a deck of playing cards?
- How many bowling pins do you have to knock down to get a strike?
- How many players are on a soccer team?
- How many players are on a baseball team?
- How many players on a baseball team play in the outfield?
- How many balls are in a pool rack?
- How many strikes do you get until you are out?

- How many points do you earn for a free throw?
- How many points do you get for a touchdown?
- How many points do you get for a soccer goal?

2. Shuffle the index cards, and place them face down on the table.

Game Rules

1. Players take turns picking a card and answering the question on the card by coming up with an expression that describes the same number in terms of another object.

For example, a player gets the question "How many sides are there on a stop sign?" She answers, "Three more than the number of fingers on one hand." Players have 30 seconds to create an expression. If they can do this, they get to keep the card. If not, they place the card at the bottom of the pile.

2. When all the cards are gone, the player with the most cards is the winner.

BRAIN Stretcher

Pick another topic, and write your own problems on an index card. Solve them with crazy expressions.

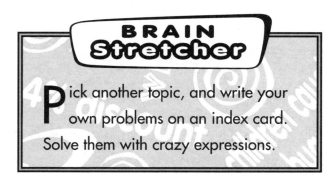

27

Just X

In many word problems, you're asked to find an unknown number. In algebra, this number is called a variable, and it is often represented by the letter x. *In this activity, you'll learn to find* x *by using cups and coins.*

MATERIALS

Magic Marker
10 index cards
4 paper cups
dish of coins
paper

Procedure

1. Use a Magic Marker to write each of the following symbols on two index cards.

$+, \ -, \ \times, \ \div, \ =$

2. Write the letter *x* on each of the four paper cups.

3. Now use the cups, the cards, and the coins to solve the following problems.

a. Ten plus a number = 35. What is the number?

Put 10 cents' worth of coins on the table. Put a + sign next to the coins. Next, put a cup next to the coins, followed by an equal sign. On the other side of the equal sign, put 35 cents. The cup, or the x, represents what you don't know. In algebra, the equation you just wrote is $10 + x = 35$.

How much money do you have to put in the cup to make both sides of the equation equal?

b. Twice a number plus 5 = 45. What is the number?

Put two cups on the table. Put a + sign next to the two cups. Put 5 cents to the right of the plus sign, and to the right of that put an equal sign. Finally, put 45 cents on the other side of the equal sign. You have to put an equal amount of money in each cup. How much money do you have to put in each cup to make the amount of money on one side of the equation equal the amount of money on the other side of the equation? In algebra, this would look like $(x \times 2) + 5 = 45$.

4. Now solve these problems on your own, using the cups.

a. Two times a number minus 3 equals 7. What is the number? (5)

b. Four times a number plus 5 equals 25. What is the number? (5)

c. Seven plus two times a number is 21. What is the number? (7)

BRAIN Stretcher

Use cups, coins, and the operation signs to make a math problem. Now change the problem you wrote into an equation.

(Answers: 25 cents, so x = 25; 20 cents, so x = 20)

Wacky Word Problem

If Mandy is 3 years older than Candy, and Mindy is 5 years younger than Mandy, who is older, Candy or Mindy? How much older?

(Answer: Candy is 2 years older than Mindy.)

28

Shoebox Math

In algebra, a mathematical sentence is an equation or an inequality. In this game of changing money, you'll learn to write and interpret mathematical sentences.

MATERIALS

pencils
paper
shoebox
bowl of coins
2 players

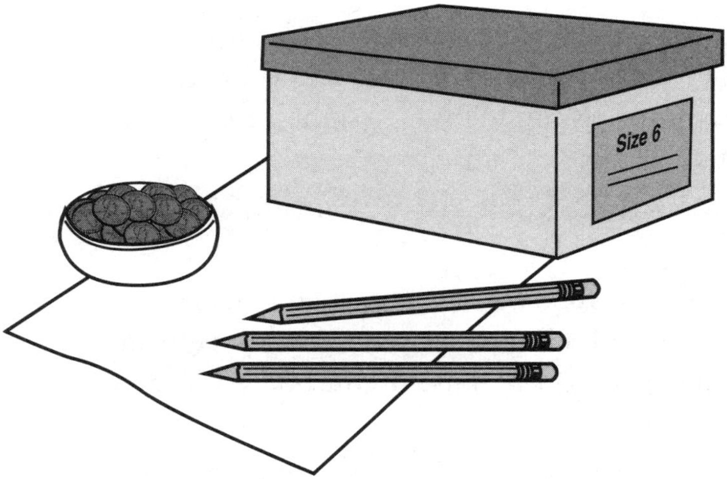

Procedure

1. Player 1 writes a simple mathematical expression on a piece of paper (for example, $x + 2$, $2x + 1$, $4x - 1$, $3x$). Player 1 folds the piece of paper, places it in the shoebox, and passes the box to Player 2.
2. Player 2 puts from 0 to 25 cents in the box (without looking at the expression) and passes the box back to Player 1.
3. Player 1 changes the amount of money in the box according to the expression (substituting the amount of money for the value of x) and passes it back to Player 2. For example, Player 1 writes the expression

$2x + 1$. Player 2 puts 8 cents in the box. Player 1 uses the original expression to calculate the new amount and puts that amount in the box. Since $(2 \times 8) + 1 = 17$, Player 1 puts 17 cents in the box.

4. Player 2 takes the money out of the box, counts the money, and writes down the result.
5. From the amount Player 2 originally put in the box and the new amount, Player 2 guesses the expression in the box. If the guess is correct, Player 2 earns 3 points.
6. If the guess is incorrect, Player 2 puts a second amount of money in the box and passes it to Player 1.
7. Player 1 again alters the amount of money in the box according to the expression in the box.
8. Player 2 takes the money out of the box, counts it, and writes down the result. Player 2 guesses the expression in the box. If the guess is correct this time, Player 2 earns 2 points.
9. Repeat Steps 5, 6, and 7, except this time Player 2 earns 1 point for guessing correctly.
10. After Player 2 guesses the expression correctly, or else doesn't guess it correctly after three tries, the roles reverse. Player 2 writes a mathematical expression on a piece of paper, folds it, puts it in the box, and Player 1 tries to guess the expression.
11. After three rounds of play, the player with the most points wins the game.

29

X and Y

In some word problems, you'll be asked to find two unknown numbers. In this activity, you'll use cups, plates, and coins to solve problems that have two variables.

MATERIALS

Magic Marker
4 paper cups
4 paper plates
10 index cards
bowl of coins

Procedure

1. Using the Magic Marker, write a large x on each of the cups and a large y on each of the plates.

2. Write the following symbols on the index cards, one on each card.

+, +, −, −, ×, ×, /, /, =, =

3. Now use the cups, the plates, the index cards, and the pennies to illustrate and solve the following problem, which has two unknown numbers. Use the cups to represent one unknown (x) and the plates to represent the other (y).

Problem: Twice one number equals a second number, and the first number plus the second number equals 15. What are the two numbers?

For the first part of the problem, put two cups on the table. Put an addition symbol between the two cups. Next to the cups place an equal sign and then a paper plate.

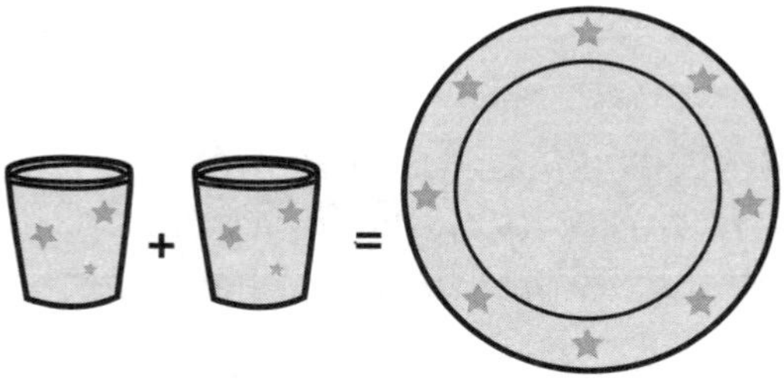

For the second part of the problem, put a cup, a plus sign, then a plate on the table underneath the first sentence. Next to the plate, put an equal sign, then 15 cents.

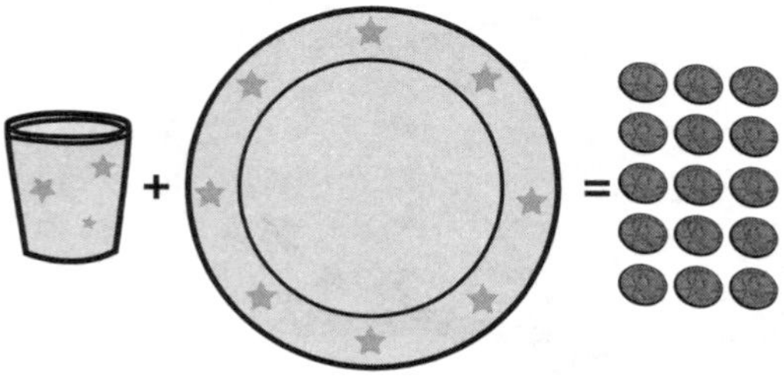

The trick to solving these problems is to change the equation so that you have all cups or all plates. Since two cups is equal to one plate, substitute two cups for the plate in the second sentence.

Now the second equation looks like this:

or $3x = 15$.

How many coins do you have to put in each cup to make the sentence true? (5)

Now that you know what a cup equals, what does a plate equal? If $x + x = y$, then $y = 5 + 5 = 10$.

4. Now try this problem on your own, using the cups and the plates for help.

Problem: One number minus a second number is 6, and four times the second number is equal to the first number. What are the two numbers? (8 and 2)

BRAIN Stretcher

Try solving this problem using the cups and the plates: Two times the first number minus the second number is 5, and the first number minus the second number is 1. ($2x - y = 5$, $x - y = 1$). What are the two numbers?

Wacky Word Problem

Seth is 2 years older than Beth. If you add their ages together, you get 30. How old are Seth and Beth?

(Answer: Seth is 16, Beth is 14.)

30

Guess and Check

Learn to solve problems by trial and error.

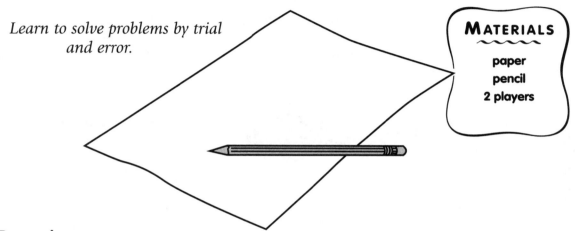

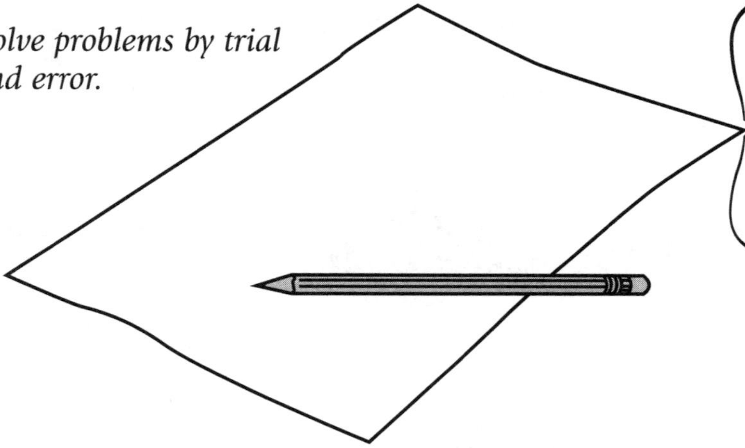

MATERIALS

paper
pencil
2 players

Procedure

1. Player 1 reads the following story with the missing blanks.

The students at Central Middle School sold ____ tickets for an 8 P.M. cheerleading competition. Adult tickets were ____ each. Children's tickets were ____ each. If a total of ______ was collected, how many adult tickets and children's tickets were sold?

2. Player 1 then fills in the blanks with logical numbers. For example:

The students at Central Middle School sold 200 tickets for an 8 P.M. cheerleading competition. Adult tickets were $5 each. Children's tickets were $2 each. If a total of $850 was collected, how many adult tickets and children's tickets were sold?

3. Now Player 2 tries to guess how many adult tickets and how many children's tickets were sold and checks the answer.

First, try any two numbers whose sum is 200.

Start with 100 adult tickets and 100 children's tickets.

These tickets cost $(100 \times 5) + (100 \times 2) =$ for a total of \$700.

Since \$700 is less than \$850, more adult tickets must have been sold.

Try 125 adult tickets and 75 children's tickets.

These tickets cost $(125 \times 5) + (75 \times 2)$ for a total of \$775.

Now try 150 adults and 50 children's tickets. What happens?

4. Now Player 2 fills in the blanks, and Player 1 tries to solve the problem using guess and check.

BRAIN Stretcher

Make up your own word problems with fill-in-the-blank numbers, and play guess and check with a friend to figure out the problems.

31

Envelope Stuffing

Stuffing envelopes isn't a very exciting job, but if you're fast, you can earn some money doing it. Let's see how fast you are. Solve some word problems to find out how much you can make in different amounts of time.

MATERIALS

- pen or pencil
- 2 sheets of paper
- 2 mailing labels
- 2 envelopes
- 2 stamps
- stopwatch
- 2 players

Procedure

1. Player 1 writes a letter or a note on a sheet of paper and writes someone's address on a mailing label.
2. Player 1 sits at a table, on which there is an envelope, the letter, the label, and a stamp.
3. Using a stopwatch, Player 2 times how long it takes Player 1 to stuff the envelope. Stuffing the envelope means Player 1 must complete all five of these tasks:
   - fold the letter
   - put the letter in the envelope
   - lick the envelope shut
   - put the label on the envelope
   - put a stamp on the envelope

   How many seconds did it take Player 1 to stuff a single envelope?
   How many envelopes could Player 1 stuff in a minute at this pace?

4. Now, using a stopwatch, Player 1 times how long it takes Player 2 to stuff a single envelope.

How long was it?

Now that you know how long it takes for each player to stuff an envelope, try to answer the following word problems.

- If each of you stuffed 500 envelopes for 5 cents per envelope, how much would you each earn?
- How long would it take each of you to stuff the 500 envelopes?
- If you and your friend were each paid 5 cents per envelope to stuff envelopes, how much would you each make if you worked for 1 hour?
- If you and your friend each wanted to earn $20, and you were each paid 5 cents an envelope, how long would it take each of you?

BRAIN Stretcher

How long would it take you and your friend to stuff 1,000 envelopes if you were both stuffing the envelopes at the same time?

Tips and Tricks

To compute how much you would make stuffing envelopes, use the following formula:

Money Earned = (price paid per envelope) times (number of envelopes stuffed)

Example

If you were paid 3 cents an envelope and you stuffed 50 envelopes, you would earn $1.50, since (.03) times (50) = 1.50.

To compute how long it would take to stuff a certain number of envelopes, use this formula:

Time = (time to stuff one envelope) times (number of envelopes to be stuffed)

Example

How long would it take to stuff 100 envelopes if you could stuff a single envelope in 10 seconds?

(10 seconds per envelope) times (100 envelopes) = 1,000 seconds

Now change 1,000 seconds to minutes by dividing by 60.

1,000 divided by 60 = 16⅔ minutes

To compute how much you would earn in an hour stuffing envelopes, first figure out how many envelopes you could stuff in an hour. Multiply the answer by the price paid per envelope.

If it took you 10 seconds to stuff a single envelope, and you were paid 8 cents an envelope, how much would you make per hour?

First, figure out how many seconds are in an hour. There are 60 seconds in a minute and 60 minutes in an hour, so there are 60 times 60 or a total of 3,600 seconds in an hour. Divide 3,600 by 10, and you will find out that you can stuff 360 envelopes in an hour. Now, if you were paid 8 cents an envelope, you would earn 360 times .08 or $28.80. Not bad.

If you want to figure out how long it would take you to earn a certain amount of money, just divide the amount of money you want to make by the amount you are paid per envelope. That will tell you how many envelopes you need to stuff. Next, multiply the number of envelopes you have to stuff by how long it takes you to stuff one envelope.

Example

You want to make $50. You can stuff one envelope in 6 seconds, and you are paid 4 cents an envelope.

Divide the amount you want to make by the amount you are paid per envelope.

$50 divided by .04 = 1,250. You need to stuff 1,250 envelopes.

Next, multiply 1,250 times 6 seconds to find the total time.

1,250 times 6 = 7,500 seconds

Divide the answer by 60 to find out how many minutes it will take you.

7,500 divided by 60 is 125 minutes, which is 2 hours and 5 minutes.

VIII

Geometry and Graphing

Geometry is the study of two- and three-dimensional space. Geometry word problems most commonly deal with finding the perimeter, the circumference, and the area. Perimeter is the distance around a shape, while circumference is the distance around a circle. Area refers to the space inside a shape. All of these measurements have a unique relationship to each other.

In this section, you will learn to figure out the area of any shape by using a piece of graph paper. You will create shape illusions, deal with some tricky triangles, play a true-false game about circles, and trade graphs in a fast-paced game of creating and solving graphing word problems.

32 Graph Paper Geometry

Use graph paper to solve common word problems about rectangles. Remember, a rectangle is a four-sided figure with two sets of parallel sides and four right angles.

Materials

graph paper
colored pencils

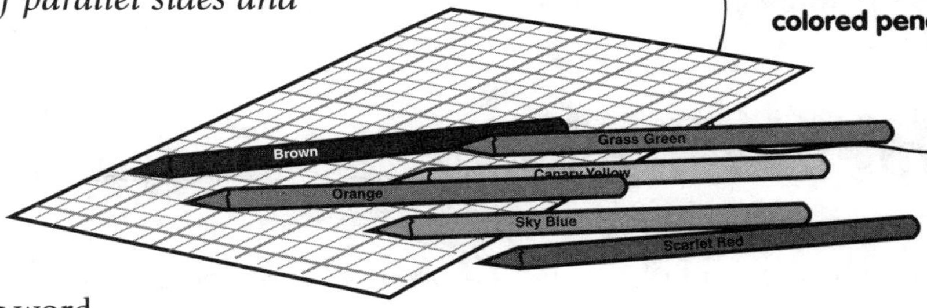

Procedure

1. Use a sheet of graph paper to solve the following word problems involving rectangles. To solve the problems, draw the sides of the rectangles on the lines of the graph paper. Count the small squares inside your figure to find the area of the rectangle. For example, a rectangle with sides of 2 units and 4 units would have an area of 8 units, and would look like the diagram shown at right.

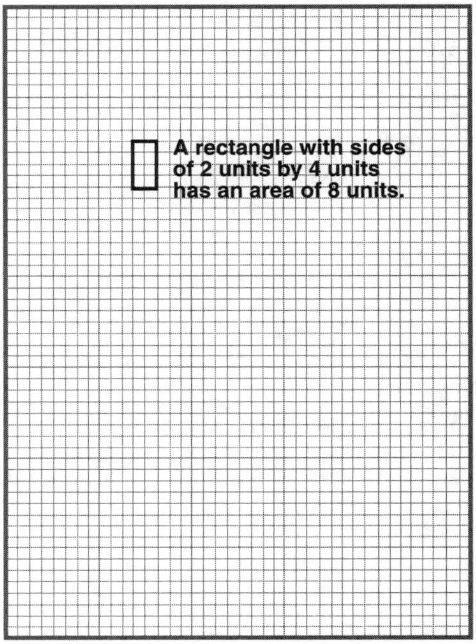

A rectangle with sides of 2 units by 4 units has an area of 8 units.

   a. If a rectangle has one side of 6 units and one side of 8 units, what is the area of the rectangle?

   Draw a side of 6 units long, and another side of 8 units long. Complete the rectangle. Count the number of small squares inside. This is the area.

   b. If one side of a rectangle is 2 units long and the area of the rectangle is 24 square units, what is the length of the other sides of the rectangle?

   c. If a rectangle has sides of 1 unit and 3 units, what is the area of a rectangle with sides that are three times as long?

2. Now try to solve the following word problems. Use the graph paper for help.

   a. The length of a rectangle is twice the width. The perimeter is 18 units. What is the area?

   b. The area of a rectangle is 10 square units. The width of the rectangle is 2 units. What is the length?

Tips and Tricks

The formula for calculating the area of a rectangle is $A = lw$, where l is the length and w is the width. Try solving each of the previous problems with the formula instead of using the graph paper, and compare the results to the answers you got using graph paper.

Wacky Word Problem

The art students want to use their own colorful tiles in one of the school's hallways. How many tiles, which are each 1 square foot in area, will it take to tile a hallway that is 3 feet wide and 20 feet long?

(Answer: 60 tiles)

33

Shape Illusions

Learn to solve perimeter and area problems while creating shape illusions.

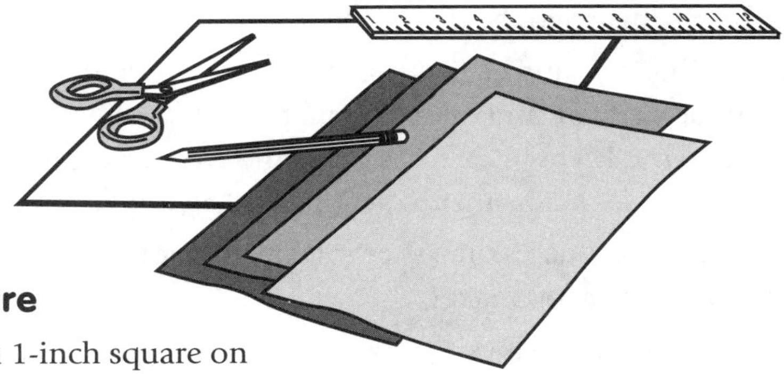

Materials

ruler
pencil
piece of cardboard or stiff paper
construction paper in a variety of colors
scissors

Procedure

1. Draw a 1-inch square on the cardboard or the stiff paper.
2. On a sheet of construction paper, make different shapes using the four squares. Create as many different shapes as you can. Trace the shapes with a pencil.

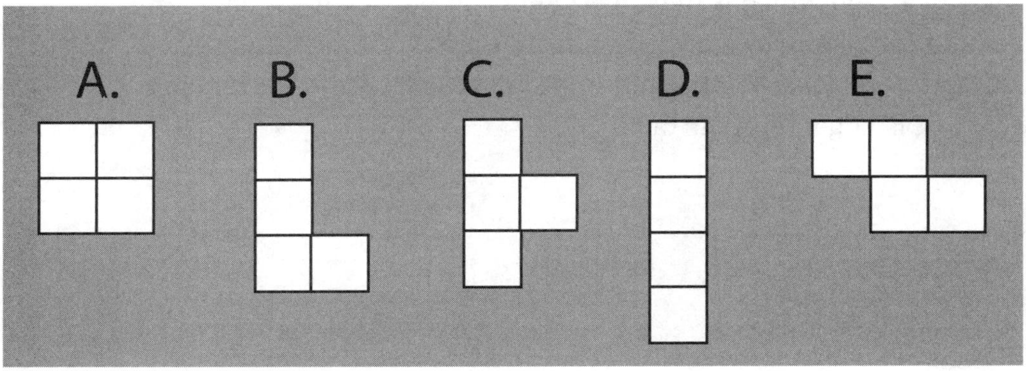

3. Cut out each of the shapes. Turn them over so that the lines you drew to create them are invisible.
4. All of these shapes have the same area, 4 square inches. What is the perimeter of each of these shapes? Use your ruler to measure each of the sides. Notice that all the perimeters are not the same. Which shape has the largest perimeter? Which shape has the smallest perimeter?
5. Each of the created shapes has an area of four. If you didn't know the area of each of the shapes, how would you figure it out? (Hint: Cut each of the shapes you created into squares and rectangles.)
   a. Shape A is a square, and Shape D is a rectangle. To find the area of these shapes, multiply the length by the width.
   b. To find the areas of Shapes B and C, first cut each shape into two parts: one part will be a square and one a rectangle. With Shape E, first cut it into two rectangles. Then, in each case, find the areas of the two smaller shapes, and add them together to get the total area.
6. Now trace shapes made of six squares on a sheet of construction paper. How many different shapes can you make?
7. Find the perimeter of each of the shapes.
8. How can you figure out the area of these shapes if you don't know that it is 6 units?
9. Now create figures out of eight squares. Cut out four of these figures, and paste them each on a separate sheet of paper. Show your four figures to a friend. Ask the friend which is the largest figure and which is the smallest. It's often hard for people to realize that the figures are all the same size. This is an optical illusion because our brain is being fooled into thinking that one shape is larger than the others.

34

Tricky Triangles

Check out the tricky triangles in this activity, and learn to solve word problems about triangles.

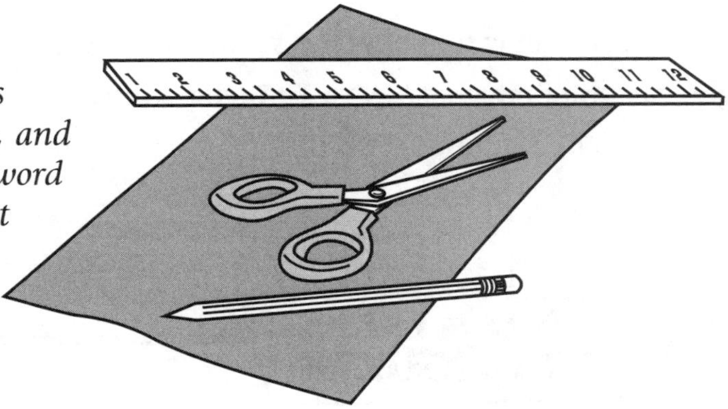

MATERIALS

pencil
construction paper
ruler
scissors

Procedure

1. Draw three different triangles on the construction paper. Each triangle should have a 4-inch base and a 3-inch height. For example,

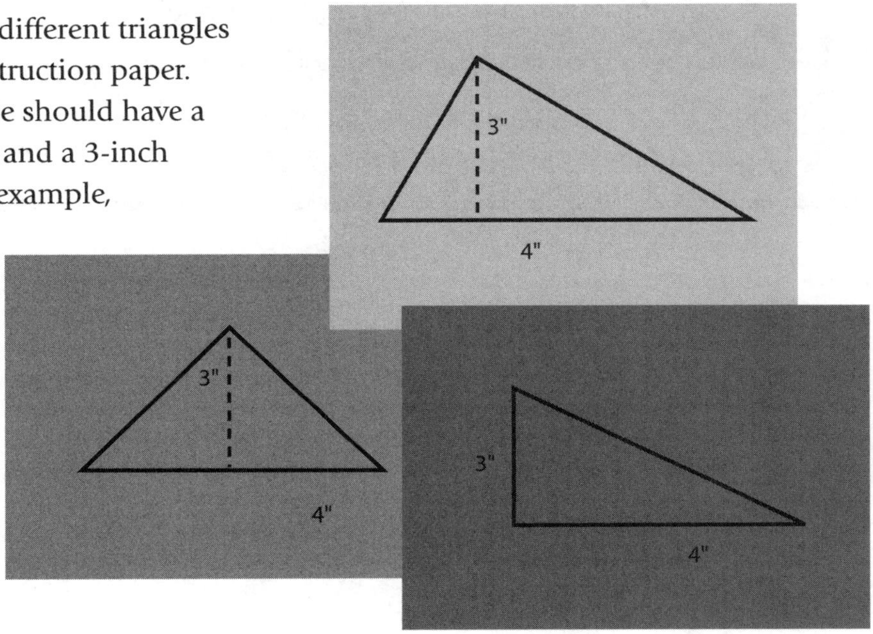

All of these triangles have the same area, which is calculated by using the formula $A = \frac{1}{2}$ (base × height).

2. Cut out all three triangles. Put them on top of each other. Do they look like they have the same area? Which one looks the biggest? Which one looks the smallest?
3. What is the perimeter of each of these triangles? Use the ruler to find out. Measure each side, and add all the sides together.
4. How many triangles can you draw that have a 6-square-inch area?
5. How many triangles can you draw that have a 12-inch perimeter?
6. Can you figure out how to make a triangle with a 12-square-inch area and a 12-inch perimeter?

BRAIN Stretcher

Draw two triangles in which the base of Triangle 2 is twice as long as the base of Triangle 1 and both triangles have the same height. Is the area of Triangle 2 twice as large as the area of Triangle 1?

35

True Circles

The circumference of a circle is the distance around the circle. It is calculated by using the formula C = π(d), *where* π *is the symbol for the number called pi (approximately 3.14) and* d *is the diameter (distance across the middle of the circle). The area of the circle is calculated by using the formula* $A = \pi r^2$, *where* r *is the radius of the circle, which is* $1/2$ *the diameter. In this activity, you'll learn to solve circle word problems while playing a true-false game.*

MATERIALS

index cards
pencils
2 or more players

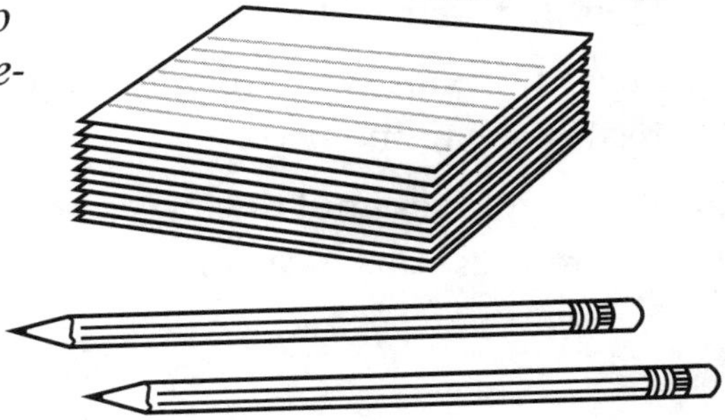

Game Preparation

1. On separate index cards, each player writes five word problems about circles, in which the answers are either true or false. Players write the correct answer, "True" or "False," under each problem.

SAMPLE QUESTIONS

True or False: A circle with a diameter of 10 has an area of 10π. FALSE

True or False: All circles with the same area have the same perimeter. TRUE

True or False: If one circle has twice the diameter of another circle, then the area of the second circle is twice the area of the first circle. FALSE

2. On the other side of the question cards, players write numbers from 1 to 5.

Game Rules

1. Players place their cards in a row, face down on the table.
2. Player 1 points to one of Player 2's cards. Player 2 reads the question on the card, and Player 1 answers "True" or "False." If the answer is correct, Player 1 gets to keep the card.
3. Player 2 points to one of Player 1's cards, and Player 1 reads the question on the card. If Player 2 answers correctly, he or she gets to keep the card.
4. Play continues, with each person picking a second and then a third card.
5. The player who got the most correct so far wins the first round. If both players answer the same number of problems correctly, no winner is awarded.
6. Players now each write three more problems to replace the three problems that were first selected.
7. Steps 3 to 6 are repeated. The player who answers the most questions correctly wins round two.
8. Play continues until one player wins two rounds. The first player to win two rounds wins the game.

Wacky Word Problem

The spring fair committee rented a hot air balloon to give rides at the fair. What is the circumference of the hot air balloon if the diameter is 20 feet?

(Answer: 62.8 feet)

36

Switch

Some word problems ask you to interpret information that is shown on graphs. Play this game with some friends and a pile of old newspapers, to create and solve word problems based on graphs.

MATERIALS

- scissors
- old newspapers
- tape
- paper
- pencils
- calculator
- table and chairs
- timer with a bell
- 3 to 6 players

Game Preparation

1. Each player cuts a graph or a chart out of a newspaper. Players can cut out pie charts, bar graphs, line graphs, or weather charts.
2. Each player tapes the chart to a sheet of paper.
3. Each player creates six word problems based on the graph or the chart and writes the questions under the chart.

EXAMPLE

This graph shows the total amount of money spent on toys each month during a 4-month period in a single household.

Look at the graph and answer the following questions:

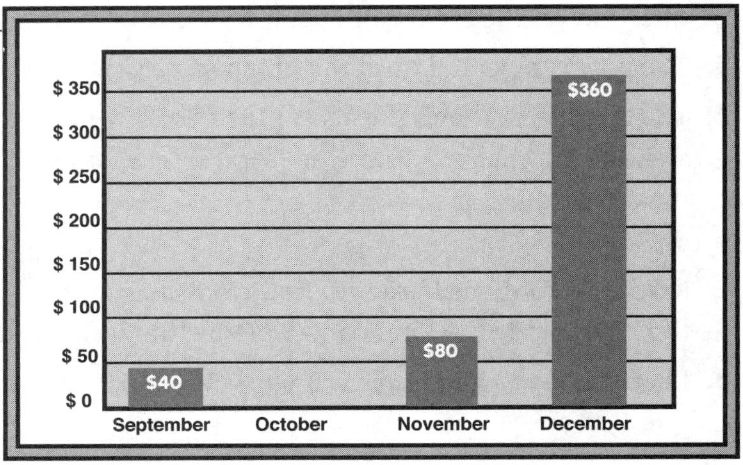

During which month was the most money spent on toys?

During which month was the least money spent on toys?

How much money was spent on toys in September?

What was the total amount spent on toys during the 4-month period?

What was the difference between toy spending in September and toy spending in December?

What was the average amount spent on toys per month over the 4-month period?

(Answers: December; October; $40; $480; $320; $120)

Game Rules

1. Players sit around a table with their charts in front of them. All of the players should have their own pencils.
2. One player is in charge of timing and sets the timer for 1 minute.
3. Players pass their graphs one player to the left. Each player has 1 minute to answer a question on the graph in front of him or her.
4. When the timer goes off, players pass the charts and the graphs one player to the left again. The timer is reset, and players have 1 minute to answer a question on the chart or the graph.
5. The play is repeated until the graphs are switched six times.
6. The graphs are then returned to the people who made them, and the answers are checked. The score for the total number correct is placed at the top of the page.
7. Were some kinds of graphs harder to interpret than others? Were some kinds of problems harder to figure out than others? Spend some extra time practicing the kinds of problems that were hardest for you.
8. Players vote on which person wrote the most creative or wackiest questions.

IX

Wrap It Up

Now that you're a word problem expert, these activities will give you more practice on all different kinds of word problems. And when it comes to math, you can never get enough practice!

Create mystery word problems using household objects, play word problem charades, and make up crazy word problems. Finally, create a set of flash cards to help you learn the most common formulas. Now that you've solved a lot of word problems of your own, it's obvious that word problems are really not so hard after all.

37

Mystery Word Problems

Create crazy word problems with household objects.

MATERIALS

2 paper bags
household objects
pencil
paper
calculator
2 players

Procedure

1. Each player takes a single paper bag, walks around the house, and places three mystery objects in it. Objects in the bag might be a fork, a book of matches, a can of soda, or anything else that will fit in the bag.

2. Players exchange bags. Each player has to write 3 word problems using the objects given. How many matches are in five books of matches? How many tines are on seven forks? How many calories are in four sodas?
3. Players exchange problems and solve them.
4. Repeat the process with three new mystery objects.
5. After four rounds, the person with the most correct answers is the winner.

38

Word Problem Charades

Practice making and solving word problems while playing a fun game of charades.

MATERIALS

pencil
paper
bowl
timer
6 players

Game Preparation

1. Divide the players into two teams of three players each.
2. Each player on each team writes a word problem on a piece of paper. The word problem should use only the numbers from 1 to 100 and should be fun to act out.
3. Each team folds up its problems and places them in a bowl.

Game Rules

1. A player on Team 1 picks a problem written by Team 2 and acts it out for his or her teammates.
2. The team has 3 minutes to figure out the problem and the answer.
3. If a team is successful in less than 1 minute, that team gets 3 points. If a team is successful in longer than 1 minute but less than 2 minutes, it gets

2 points. If the team is successful in longer than 2 minutes but less than 3 minutes, it gets 1 point.

4. A player from Team 2 picks a problem written by Team 1 and acts it out for his or her team to solve.
5. Each player takes a turn acting out a problem. After all three rounds, the team with the most points wins the game.

Wacky Word Problem

Try acting this one out: A child got 50 goldfish for her birthday. She gave 6 of those goldfish to a friend. How many goldfish did she keep?

39 Answers, Answers, and More Answers

Create word problems that fit specific answers.

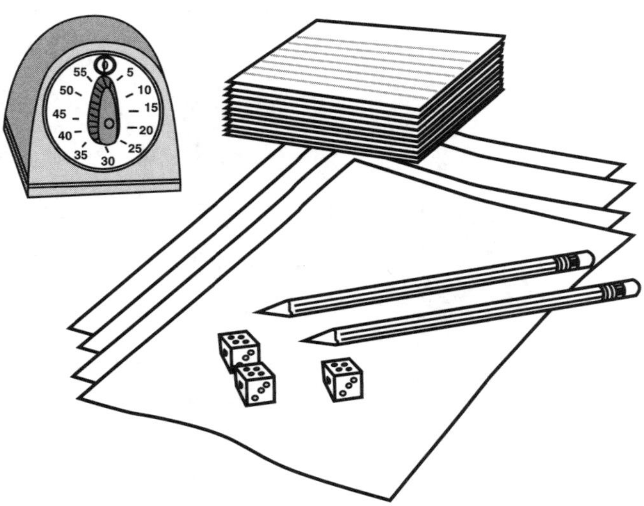

MATERIALS

2 pencils
10 index cards
3 dice
timer
several pieces of paper for each player
2 or more players

Game Preparation

Write one of the following words or phrases on each index card.

__________ miles per hour

__________ feet per minute

__________ gallons

__________ meters

__________ feet

__________ percent

__________ feet per second

__________ square feet

__________ hours

__________ grams

Game Rules

1. Shuffle the cards, and place them face down on the table.
2. One player draws a single card. Another player rolls all three dice. The number rolled is written in the blank on the card.

EXAMPLE

If "12" is rolled and "miles per hour" is drawn, the card now reads, "12 miles per hour."

3. Each player now has 3 minutes to write as many word problems as possible, for which the answer is 12 miles per hour.
4. When the time is up, each player reads his or her problems aloud.
5. The player who came up with the most problems is the winner.

40

Formulas . . . Formulas . . . Formulas

Formulas are helpful in solving many types of word problems. In this activity, you'll create a set of formula flash cards to practice with, so that you'll have the formulas in your head when you need them.

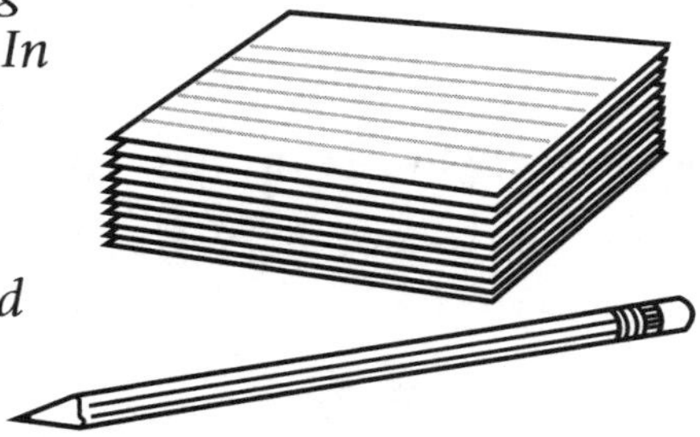

MATERIALS

index cards
pencil

Procedure

1. Write each of the following formulas on an index card, one formula per card.
2. Write the question that the formula answers on the other side of the index card.

EXAMPLE

On one side of the index card, write the question, "How do you find how far it is from one place to another?"

On the other side of the index card, write the formula $D = RT$.

(Hint: If you need more examples, look back at some of the activities in this book.)

3. Practice asking yourself the questions on one side of the card, and see if you can recite the formula on the other side of the index card.

Measurement Problems

1 kilometer = 1,000 meters

1 meter = 100 centimeters

1 meter = 1,000 millimeters

Percentage Problems

To find a certain percentage of a number = Change the percentage to a decimal and multiply by the number.

To find what percentage one number is of another number = Divide the first number by the second number and multiply by 100.

Interest Problems

Interest Earned = Initial Investment × Interest Rate × Years Invested

Distance Problems

Distance = rate × time ($D = RT$)

Time = distance ÷ rate ($T = D/R$)

Rate = distance ÷ time ($R = D/T$)

Geometry Formulas

Perimeter of a square = side + side + side + side ($s + s + s + s$)

Area of a square = side × side ($s \times s$)

Perimeter of a rectangle = length + width + length + width ($l + w + l + w$)

Area of a rectangle = length × width (lw)

Perimeter of a triangle = side + side + side ($s + s + s$)

Area of a triangle = ½ base × height ($\frac{1}{2}bh$)

Perimeter of a circle = Pi × diameter (πd)

Area of a circle = Pi × (radius × radius) (πr^2)

Word Problem Master Certificate

Now that you've mastered all of the different types of word problem–solving techniques in this book, you are officially certified as a word problem master! Make a photocopy of this certificate, write your name on the copy, and hang it on the wall.

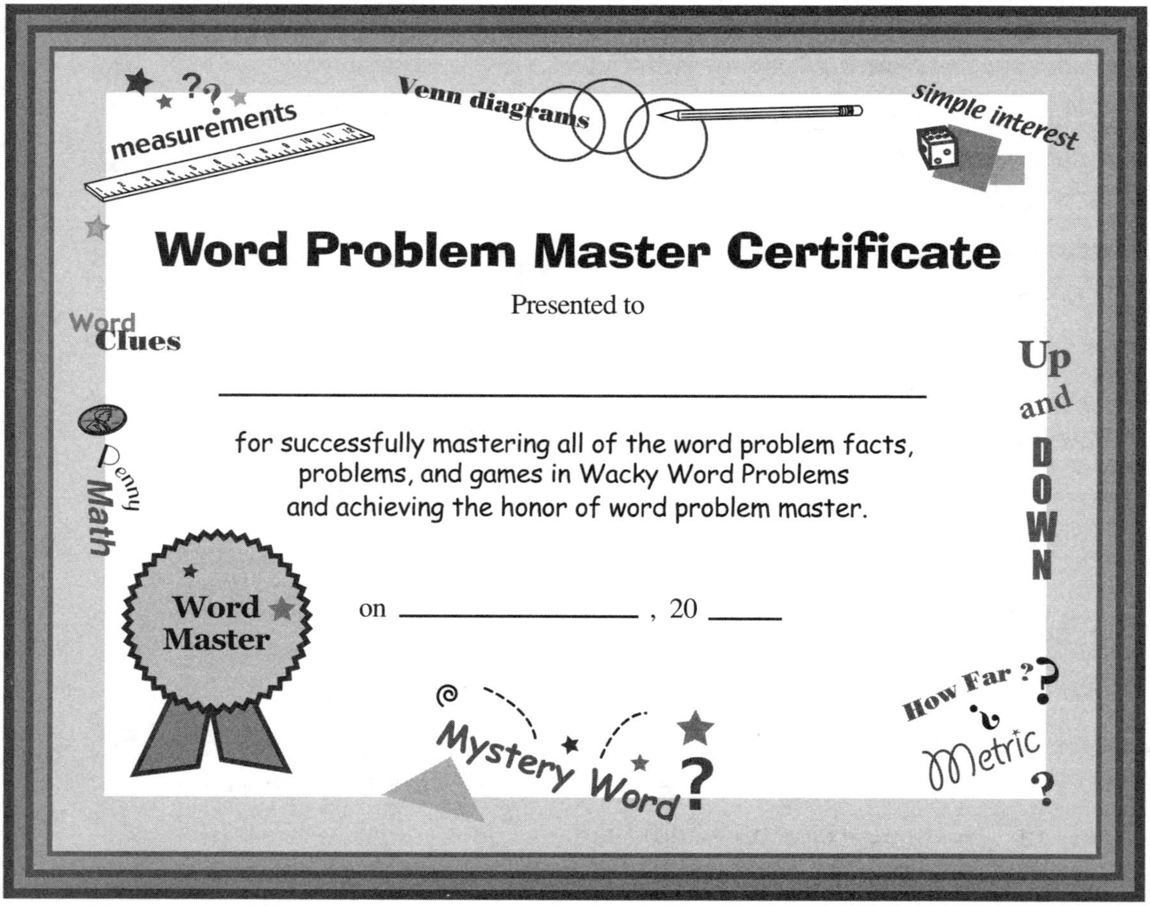
measurements
Venn diagrams
simple interest
Word Problem Master Certificate
Presented to
for successfully mastering all of the word problem facts,
problems, and games in Wacky Word Problems
and achieving the honor of word problem master.
on ______________ , 20 ____
Word Clues
Penny Math
Word Master
Up and DOWN
Mystery Word?
How Far ?
Metric

Index